NATO ASI Series

Advanced Science Institutes Series

A series presenting the results of activities sponsored by the NATO Science Committee, which aims at the dissemination of advanced scientific and technological knowledge, with a view to strengthening links between scientific communities.

The Series is published by an international board of publishers in conjunction with the NATO Scientific Affairs Division

A Life Sciences B Physics	Plenum Publishing Corporation London and New York

C Mathematical and
   Physical Sciences
D Behavioural and
   Social Sciences
E Applied Sciences
	Kluwer Academic Publishers
	Dordrecht, Boston and London

F Computer and
   Systems Sciences
G Ecological Sciences
H Cell Biology
I Global Environmental
   Change
	Springer-Verlag
	Berlin Heidelberg New York
	London Paris Tokyo Hong Kong
	Barcelona Budapest

NATO-PCO DATABASE

The electronic index to the NATO ASI Series provides full bibliographical references (with keywords and/or abstracts) to more than 30000 contributions from international scientists published in all sections of the NATO ASI Series. Access to the NATO-PCO DATABASE compiled by the NATO Publication Coordination Office is possible in two ways:

- via online FILE 128 (NATO-PCO DATABASE) hosted by ESRIN, Via Galileo Galilei, I-00044 Frascati, Italy.

- via CD-ROM "NATO Science & Technology Disk" with user-friendly retrieval software in English, French and German (© WTV GmbH and DATAWARE Technologies Inc. 1992).

The CD-ROM can be ordered through any member of the Board of Publishers or through NATO-PCO, Overijse, Belgium.

The ASI Series Books Published as a Result of
Activities of the Special Programme on
ADVANCED EDUCATIONAL TECHNOLOGY

This book contains the proceedings of a NATO Advanced Research Work-shop held within the activities of the NATO Special Programme on Advanced Educational Technology, running from 1988 to 1993 under the auspices of the NATO Science Committee.

The books published so far as a result of the activities of the Special Programme are as follows (further details are given at the end of this volume):

Item Banking:
Interactive Testing and
Self-Assessment

Edited by

Dieudonné A. Leclercq

Service de Technologie de l'Education
Université de Liège au Sart Tilman
Boulevard du Rectorat, 5 Bât. B32
Sart Tilman, B-4000 Liège, Belgium

James E. Bruno

University of California
Graduate School of Education
405 Hilgard Avenue
Los Angeles, CA 90024-1321, USA

Springer-Verlag
Berlin Heidelberg New York London Paris Tokyo
Hong Kong Barcelona Budapest
Published in cooperation with NATO Scientific Affairs Division

Proceedings of the NATO Advanced Research Workshop on Item Banking:
Interactive Testing and Self-Assessment, held in Liège, Belgium, October
27–31, 1992

CR Subject Classification (1991): K.3, H.3

ISBN 3-540-56653-8 Springer-Verlag Berlin Heidelberg New York
ISBN 0-387-56653-8 Springer-Verlag New York Berlin Heidelberg

Typesetting Camera-ready by authors
45/3140 - 5 4 3 2 1 0 - Printed on acid-free paper

Preface

The international conference held in Liège, Belgium, October 26 – 31, 1992, was intended to bring together scholars from around the world to discuss and critically analyse the issues and problems associated with subjective probability measurement or the more generic research area called self-assessment. Recent advances in computer technology (hypermedia, expert systems, interactive video disks, etc.) along with the developing sophistication of self-assessment scoring systems based on Subjective Probability Measurement (SPM) made this conference particularly important and timely. The relevance of papers presented at this conference to the enormous educational and training needs of the "new" European community is of special note.

One theme that permeated much of the discussion at the conference was the need not only to better articulate SMP methods, but to demonstrate their value in technology based self-assessment systems with regard to supporting instructional programs. The formation of a self-assessment and evaluation research organization within the context of a European Research Organization was unanimously recommended by conference participants. The University of Liège and the University of California at Los Angeles, more specifically our respective departments, have agreed to serve as a world-wide clearing house on information with regard to advances in Subjective Probability Assessment.

The reader will note not only the high quality, both theoretical and applied, of the papers presented in these proceedings but also their immediate relevance to educational issues in both formative and summative evaluations. The fact that these papers are the most up-to-date reflections of the field by world-class scholars adds to the relevance and importance of this conference.

In summary, Subjective Probability Measurement (SPM) is a difficult theoretical concept for most educators to comprehend. The sophisticated nature of modern computer systems (expert systems, interactive video disks, and hypermedia) coupled with comprehensive formative and summative evaluation and self-assessment systems make SPM transparent to the user. This transparency should increase its acceptability across various education client groups. These proceedings have been intended to enlighten as well as stimulate the reader towards further research in each of these three major components of the conference - the inputs (items and computer environment), the process (the SPM methodology), and the outputs (the actual self-assessment system of reports to support instructional programs). As developments proceed in each of these three areas, what this conference has achieved is a framework within which to place the developments. The expressed intent of the conference of improving educational formative and summative evaluation by means of SPM and self-assessment systems was clearly underscored by several papers presented in these proceedings. The reader is encouraged to pursue reading some of the support articles of the scholars presenting their research at the conference that are noted in the bibliography.

July 1993   Dieudonné A. Leclercq
                                                                   James E. Bruno

Acknowledgments

This conference has been made possible by the support of the NATO Scientific Affairs Division, the "Fonds National de la Recherche Scientifique" of Belgium, the Ministry of Education, Research and Training of the French Speaking Community of Belgium, the University of Liège, Belgium, and the Saffraanberg Technical School of the Belgian Air Force.

The following persons contributed in making this conference a success: Prof. Véronique De Keyser, Dean of the Faculty of Psychology and Education, Liège University, who welcomed the participants and Prof. emeritus Gilbert De Landsheere, World Award of Education, who gave a final address.

We would like to thank the following members of the Service de Technologie de l'Education, Liège University: Jean-Marie Lemaire, Michel Jacques, Robert Carion and Nelly Saenen, and especially Ingeborg Frank, who typed the final version of the proceedings and helped during the pre, per, and post phases of the conference, and Jean-Luc Gilles, who operationally and administratively prepared and orchestrated the whole conference and made things really happen.

ontents

Editor's Introduction

He who knows and knows that he knows is wise - follow him
He who knows not and knows not that he knows not is a fool - shun him
He who knows not and knows that he knows not is a child - teach him
He who knows and knows not that he knows is asleep - awaken him

Arab proverb

Pues amarga la verdad quiero echarla de la boca[1]

For decades, assessment has been recognized as a key feature in learning efficacy, especially through formative evaluation. But, for decades too, users have been deceived by educational assessments from many respects.

As early as 1972, the Centre for the Study of Evaluation (CSE) and Research for Better School (RBS) analyzed 2,600 tests on higher order cognitive, affective and interpersonal skills and competencies. The criteria used for their analysis organized them according the VENTUR acronym: V for validity, E for Examinee appropriateness, N for Normed excellence, T for Teaching feedback, U for Usability and R for Retest potential. Summing up dozens of criteria, evaluations of P (Poor), F (Fair) or G (Good) were attributed to each test representing its ratings value from each of the 6 standpoints.

The modal pattern for higher order cognitive skills tests appeared to be PGPPGP: all poor except Examinee appropriateness and Usability. In other words, educational tests "fit" the school system but do not mean anything (poor validity), can be used neither in a normative perspective (poor Norms), nor in a formative one (poor Teaching feedback) and cannot be used to measure gains (poor Retest potential).

Twenty years later, about the same general criticisms can be directed to current educational assessments. They largely suffer from **validity problems**: they often address the lowest level of Bloom's cognitive taxonomy (rote memory). Consequently, they hardly test critical reasoning, cognitive vigilance, or capacity to analyse and make decisions.

Enormous **reliability problems** exist as well: correction is not objective, as has largely been demonstrated by the "docimology" movement started by Pieron (1963) in France.

---

[1] Since truth is bitter, I want to spit it out of my mouth. From the Spanish poet Francisco de Quevedo.

Sensivity problems are obvious: the measure of knowledge is insufficiently subtle, students are too scarcely, if ever, invited to express how partial or total they consider their knowledge is. The feedback itself is far too global and non-diagnostic. It is too often reduced to one score and usually does not refer to specific objectives, contents or abilities.

Communication problems are permanent: delay for feedback is often excessive and consequently loses its relevance and motivational power. Moreover, feedback is mainly reduced to logic or arithmetic modes of communications, whereas graphics, verbal or iconic facilities are very much underexploited.

Management and userfriendliness problems could not be forgotten: students are not free enough to choose the content, the moment and the way (inductive or deductive approach, with or without help) in which they want to be tested.

Finally, **research problems** should be stressed: whereas they could be a great source of information, practical data from actual settings are mostly insufficiently processed to be of benefit to fundamental research. In addition, the dialog between cognitive theories and practical applications, i.e., between laboratories and classrooms, is insufficient.

Edumetrics: an explosion of solutions

Coined by Carner (1974), the word edumetrics can be used to express the three decades of efforts made by educational research in trying to answer some of the aforementioned drawbacks and weaknesses of educational testing. Examples of this endeavour are latent trait theories (and the Rasch Model), generalisability theory, tailored testing, placement tests and individually prescribed instruction, Intelligent Tutoring Systems (combining Computer Assisted Testing and Artificial Intelligence).

Among those efforts, special attention should be given to confidence marking. Starting in the early 1960s, a stream of studies have been dedicated to scoring multiple choice questions according to self-assessment estimated by the learner. Unfortunately, those studies contained such a number of methodological flaws (see Leclercq in this volume) that even scholars such as Shuford (with his 1966 key paper) did not receive the attention deserved and that the whole domain was considered, from the mid-1970s, as a gigantic dead end.

Fortunately, scientists like Shuford, Hunt, and Bruno in the United States and De Finetti, Van Naersen, Dirkzwager, Van Lenthe, Fabre, and Leclercq in Europe have continued to explore the area, proving that not only it is not dead, but that is has, after a dozen years of apparent "jachère" (fallowness) become a vivid movement where methodological constraints are fully recognised and overcome.

It is our hope that this conference will contribute to promoting this "Self-Assessment" movement on solid ground.

When these features have become more familiar, links will be elaborated between Self-Assessment and the Rash Model (since both are deeply concerned with the probability of a given person answering correctly a given item) as well as generalisability theory (since both aim to assess the reliability of a student's measure of knowledge).

The reader can now understand why self-assessment and subjective probability specialists provide a central contribution to this volume, referring to De Finetti's (1965) words: "Only subjective assessment can contribute to objective measurement of knowledge".

A second and more recent stream should also be considered with attention: improvements to educational testing due to *ad hoc* use of computers. In this respect, the **hypernavigational** approach represents an alternative to the teacher-directed "dialog", to "one-way communications", and, as will be seen, opens perspectives to new modes of evaluation, especially of self-assessment.

In the same way, since the recent availability of video, sound, iconic and graphic resources, **multimedia** developments and theories change the way the problems of evaluation are stated and, of course, the types of answers that could be given, in a realistic way (i.e., at hand for average users).

As a consequence, **item banking** principles provide the general background, justifying the social utility of the whole concern and nationwide perspectives. They also will elucidate the need for appropriate types of questions and modes of processing the answers.

This variety of concerns may appear as a Babelian construction. We hope that the following papers will show how far the kinds of expertise provided match with the listed problems.

The research papers found in these proceedings are organised as follows:

Inputs	Process	Outputs
Item banking technologies such as expert systems hypermedia, etc., and item banking itself	Subjective Probability Measurement (SPM) developments and emphasis on SPM scoring system	Self-assessment procedures that produce various evaluation reports for instructional support systems

New Technology Implementation:
Item Banking in Holland

E.C.M. Verwaijen

Innovation Centre for Vocational Training and Industry (CIBB), Pettelaarpark 1 's-Hertogenbosch, Postadres, Postbus 1585, 5200 BP 's-Hertogenbosch, Holland

Abstract: In Holland all official examinations are organized by central national training organizations. These exams are summative evaluations.

Modular systems and flexibilization are in full swing. At the same time, the various training institutions tend to become too large as a result of mergers. This means more different examinations have to be produced, that there must be more frequent times of examination and more types of examination. In practice, the feasibility of "tailored" flexible training also depends on a flexible and controllable test and examination system. So the introduction of a computerized item bank system seems much desired.

This paper contains a report of a research on existing or developing item bank systems in Holland. The CIBB was commissioned by the common national training organizations. Part of the commission was coming up with a well-founded implementation proposal.

1 Introduction

CIBB stands for Centrum van Innovatie voor het Beroepsonderwijs en het Bedrijfsleven (Innovation Centre for Vocational Training and Industry). So the sector directly served by the CIBB is the vocational training sector, in its relation with industry, in all its aspects.

Vocational training in Holland and its organization is as follows: training institutions (technical and vocational training for 12-16 years old, for 16-18 years old, and for 18+) and learning while you work (the apprentice system). In the apprentice system the pupils or students attend school for one day, and work at the (apprenticeship) firm the other four days of the week. We are concerned with primary, secondary and tertiary types of training.

The various branches have each their own foundation to support the specific branch-oriented type of training. They are the national training organizations. Some of them are the Foundation for Vocational Training for the Metal Industry, the Foundation for Vocational Training for the Installation Technology branch, etc.

The clients of CIBB are schools and the national training organizations, as well as umbrella organizations of schools and national training organizations. Besides these there are the employers' and employees' organizations, the various ministries (e.g. Education and Science), the Employment Exchange and institutes, organizations and companies that are in any way involved with or related to training.

2 Implementation

It is essential in the case of computerization and automation projects that the users are involved in the entire process. Implementation starts as early as preliminary research.

Many projects of this type fail because users are too often left out. The educational sector is no exception. Not they, I might add. In my experience I have often noticed resistance against computers.

Teachers and assistants being afraid and ignorant of anything having to do with computers, and having bulging agendas as well, we have to include a lot of time in our projects for implementational guidance. Not one project is therefore complete without well-founded suggestions for implementation.

I do a lot of project concerning automatisation of information and every time I realise the importance of guiding the users. In the project "taking examinations by computer" I spend more time guiding the users and making recomandations of using an itembank than in the reseach itself.

3 Project "Taking examinations by computer"

In these days of computerization people and things alike tend to be computerized. It is difficult not to become a victim of this proliferation. Computerization is the name of the game, so computerization it will be. Which reminds me of a Dutch cabaret performer saying: "I've bought a car, it's a great help when looking for a parking place".

Sometimes, however, you cannot get around opting for computerization as a solution. In Holland modular systems and flexibilization are in full swing. At the same time, the various training institutions tend to become too large as a result of mergers. Also, the view of learning and taking exams is changing as will become clear during this conference. More concretely this means that more different examinations have to be produced, that there must be more frequent times of examination and more types of examination. In practice, the feasibility of "tailored" flexible training also depends on a flexible and controllable test and examination system.

The CIBB was commissioned by COLO, the common national training organizations, to do research on existing or developing item bank systems. There are so many of them: you cannot see the wood for the trees, which is the best, how do I make a selection.

Part of the commission was coming up with a well-founded implementation proposal in the form of recommendations: a stepped plan.

4 Construction: National training organizations versus the taking of examinations

National training organizations are responsible for all official examinations taken at vocational training schools for 12-16 years and 16-18 years for their specific branches. There are item authors, examination producers, examination inspectors, all within their specific branches.

Looking at the developments in the field of modular systems and flexibilization outlined above, you will understand that this is becoming more than one can handle. The development of good examinations requires a great effort. Especially the production of good items takes a lot of time.

We did some research to determinate how many items on paper can be used in an itembanksystem and we came to calculate that only 20% of the already existing items can be used. So it really takes a lot of time to constitute an itembank.

The introduction of a computerized item bank system seems much desired. Only, this has or had to be done side by side with the current way of producing examinations.

A lot of time will be spent into item and matrix development, learning how to deal with a selected package, with examinations being taken as usual. This clearly calls for a multi-year planning.

Another important aspect is the large difference in size and organization of the various training organizations. For instance, we visited training organizations with different functionaries for the item bank, its development and its implementation. But we also visited training organizations where the examination coordinator was responsible for the production of items as well as the production of examinations and their distribution. There is no doubt that this functionary will lack time on all fronts.

5 Research subject, phasing

The research had been organized as follows:
. literature study on the item bank systems in use or under development in Holland. The result was a report describing the various item bank systems.
. problem analysis: what about the examination construction. What is the problem. What is wished for. What type of equipment is used. How are institutions organized. Are there any other systems to which an item bank system should be linked. This resulted in a report that clearly outlined the limits and contours of an item bank system.
. the creation of an ideal-typical item bank system. We translated our findings into a matrix with general and specific wishes/characteristics.

The standard for the general characteristics was as follows: fundamental requirements, general requirements and characteristics of 90% or more of the interviewees. Specific characteristics were those that were found only at one or two individual national organizations. Taking stock of the characteristics we found that when a characteristic proved not to be general, it had indeed been found at only one or two organizations.

6 Major requirements

The following were found to be major requirements:

Some fundamental requirements, besides obvious requirements of storage and production of items as well as matrices and tests, were:
- the production of the tests must be possible with the aid of matrices, which can also be stored in the system
- for quality control and maintenance of the item bank, psychometric data are indispensable
- the system must have a structure with at least 3 (hierarchical) layers and allow of simply being extended (extension of structure).

The majority of requirements mentioned regarding the item data that must allow of being stored are the following:
- type of question (multiple choice, open)
- correct answer
- at least four alternatives to a multiple choice question
- date of entry into the bank
- date of retrieval (for use in an examination)
- reference to source or picture
- psychometric data
- key to classify the subject matter
- status of item (new, in use or being repaired, evaluated)

In addition to these categories of requirements, another number of characteristics of the item bank packages was added to the checklist, as well as some general requirements such as:
- required minimum hardware configuration
- package capacity (maximum number of items)
- costs
- supplier's support
- safeguarding provisions
- export to a word processor (WP)

7 Results

The three reports resulted in a selection of five item bank systems. The other item bank systems in no way came up to the criteria, limits and/or contours. These other itembanks are discribed in the report.

The five item bank systems selected were scored (in cooperation with the developers) on the list of characteristics (the major requirements).

One of the objects of the eventual report was a merging of all preceding reports, with the scored list of characteristics being the most important part. Educational aspects were studied side by side with computerization aspects. The result was a sort of quality test with a subsequent value judgment.

The five itembank systems:

TSSA

Developed under supervision of Mr. A. Horsten of Kathotic University Nijmegen. The itembank is not very expensive. It is a simple system wich can be used by training institutions with a simple organizational structure. This means that the data structure is closed within an individual subject. It requires minimum hardware configuration. The package capacity (maximum number of items) is 1100. There is no export to a word processor (WP).

Mastertool

A commercial product. The production of the tests is possible with the aid of matrices, which can also be stored in the system. The itembank is expensive. There is supplier's support, safeguarding provisions, export to a word processor (WP). They sell on the basis of "your wish is our command" : no balance options against each other and relate them to those educational starting points. Users can select psychometric data.

Toetshulp (CITO)

You can not use Toetshulp without the help of CITO, the examination-organisation. They develop and produce the questions and the matrices. You use the itembank filled by CITO.

The Examiner

A commercial product. The itembank is expensive. There is a great supplier's support, safeguarding provisions, export to a word processor (WP). They sell on the basis of "your wish is our command": no balance options against each other and relate them to those educational starting points. Only the p-value is processed. Users can select questions by given p-value.

Microbank

Microbank, which was developed under the supervision of Mr. E. van Hees of Brabant University. The majority of the national training organizations have meanwhile purchased or intend to purchase this package. It can be used by organizations with a intricate organizational structure. The production of the tests is possible with the aid of matrices, which can also be stored in the system. Psychometric data are stored for quality control and maintenance of the item bank. The system has a structure with 5 (relational) layers and allows easy extension of structure.

8 Implementations suggestions

Apart from a sensible choice, the implementation requires a great deal of attention.

1. The choice and introduction of an item bank system is an educational as well as a computerization project.
A sufficient amount of time needs to be reserved for classification, item screening, production of matrices. Experience warns against an over-optimistic idea of the usability of existing components like matrices and items.The difference in size and organization of the various training organizations is such that the purchase of an item bank package is bound to fail without the help of a stepped plan.
The choice of an item bank package and the purchase of a computer mark the beginning of a long road. Before examinations can be produced, the item bank must be fed with information. This is done in a certain order, which requires a classification of all training matter, split up into progressively smaller components. These components have to be stored cohesively.

2. The tasks of several of the people involved in the examination production will become different. An important point here is that the quality of the examinations produced with a computerized item bank system is proportional to the quality of the employed classification structure and the items and matrices that have been developed.

3. The implementation of an item bank system is best viewed as a project. The project leadership is best undertaken by somebody having expert knowledge, authority and vision in the field of item banks as well as renovation processes: both an opinion leader and a change agent.

9 A sensible choice

There is no item bank system that makes all the wishes of the national training organizations come true.
The choice of an item bank system depends on the organization, the method of examination, the size of the institute, etc.
The following are some questions that an organization should ask itself:
- Contracting out or doing the job in-house? There are organizations that take the entire examination production and organization off one's hands.
- Is computerization the best solution? Sometimes the best thing is first to properly organize the examination production manually.
- Are several people supposed to start working with the item bank system at the same time?
- Is the examination production divided over various banks?

So far we have led ourselves to be guided by the educational starting points. While commercial products sell on the basis of "your wish is our command", a reasoning educational system will balance options against each other and relate them to those educational starting points. Think e.g. of the possibility whether or not to be able to select for psychometric data, automatic removal of an item with too high or too low a P-value.

Other important aspects connected with a sensible choice derive from an analysis of the organization and the data structure of the composition of the examinations.

Training institutions with a simple organizational structure mostly keep their subjects separate, have separate functionaries for the various subjects and have subject-related examinations. This means that the data structure is also closed within an individual subject. This is, for instance, the case with secondary schools. Subjects like Dutch and history are taught separately, by separate teachers, while it will be impossible to use questions of one subject on behalf of the other for examination purposes.

The vocational training institutes have a rather more intricate organizational structure. Subject tend to blend, flexibilization and modular systems are key-words. Also, the examination organization is more complicated. There is, for instance, an overlap between various subjects: between mathematics and electrical engineering. Examinations use each other's questions.

It also happens that subjects have general subject matter, e.g. "colour study" with the graphical industry, and specific subject matter like "patterning", which, again, uses colour study.

10 Additional implementational support

The CIBB currently also supports the various organizations in purchasing and introducing the item bank. For instance, we provide general guidance in writing items. But we also give courses in all important fields.

I hope I have succeeded in giving you an idea of the situation in Holland in respect of item banks in Vocational Training.

Particularly, I have tried to make you see how important the implementation of new technologies is. We may devise beautiful things or make recommendations to the educational field, but if we forget about implementation or make a poor job of its preparation or realization, any new introduction is bound to fail.

An Item Banking Service:
Pre-Project for a National System of Evaluation Tools

Umberto Margiotta [1] and Renata Picco [2]

[1] European Centre for Education (CEDE), Ca' Foscari, University of Venice, Chair of Pedagogy and Theory of Instruction.Villa Falconieri, 00044 Frascati (Roma), Italy
[2] European Centre for Education (CEDE), Computer Department. Villa Falconieri, 00044 Frascati (Roma), Italy

Abstract: In the framework of CEDE strategies in educational research work, the opportunity was considered to establish an Item Banking System. The system aims at promoting and coordinating the development of measurement, regulation and control tools and at disseminating their use, in a variety of institutional educational settings, as well as in different social contexts, where educational actions are required and particular attitudes towards self-assessment are the mainstream. The targets of the service, generally speaking, are people working in situations in any way involved in processes of enlarged social reproduction and social change. CEDE's National Item Bank is intended to operate as a system of evaluation tools which, as a consequence of a continuing inquiry into the existing needs, pursues user-defined educational goals and generally follows a user-centred orientation. It forms an integrated system for: storage and management of educational test items and related statistical data, test construction, and analysis of the results of test administration. The project implies a non-centralized storage and processing system, but a networking of resources. It should support local banks (schools, education and training agencies) as the usage and maintenance of test items and result analysis. A particularly enhanced technology will be used in order to optimize the potentialities of Item Banking. The management system and the interface module for user's access will be of 'object oriented' design.

Keywords: Evaluation system, evaluation tools, educational testing

1 Preface. The Demand for Evaluation Tools
Why an Item Bank?

In the framework of CEDE's strategies in educational research work, the need was acknowledged to establish an Item Bank Service. Reasons for this can be summarized as follows:

many research projects run by CEDE in the 1980s made use of evaluation tools (IEA surveys [1]), or worked them out anew (Vamio [2]), in order to assess school achievements and capabilities, as well as non-cognitive aspects. An internal organization was activated, functional to the implementation of these surveys, for the dissemination of tests, their return and the data processing. But great interest and expectations arose about a possible broader utilisation of such instruments, showing the likelihood of a strong demand coming from schools, institutions, individuals, for a service providing standardized but flexible tests. Teachers, project-leaders, researchers, decision-makers and trainers complain about the absence, to date, of a system allowing them to have access to reliable and user-oriented evaluation tools. Tests are required to make comparisons, for diagnostic aims, and to get feed-back information about the effectiveness of programmes. Such a system could act as an Item bank service.

[1] The CEDE is a National Centre in Italy of the International Association for the Educational Achievement (IEA, see Husen), that is to say the institution which is the Italian member of the IEA. Over the past decade, the CEDE was involved in five IEA international projects, large-scale comparative surveys of educational performance. Some of these have just begun or are ongoing; others have just completed. Per un quadro sintetico of IEA studies ongoing at CEDE in '80's, see Visalberghi (1990) and papers of Proceedings of Open Day of IEA General Assembly 1988 (Frascati, September 1988), see references: Second IEA Science Study (SISS), Fierli (1985), Fabi (1986) e Melchiori (1989); Indagine sullaProduzione Scritta (IPS), Lucisano (1988); Computers in Education (COMPED), Caputo (1990); ricerca CEDE sui livelli di alfabetizzazione funzionale dei corsisti delle 150 ore, svolta nel '91 nel quadro della IEA Reading Literacy, Lichtner (1991) e Caputo (1991); indagine IEA sull'Educazione Prescolastica (Pre-Primary), Pusci (1992).

[2] R. Bolletta proposed (1985) and carried out at CEDE, a study titled VAMIO (Verification of Mathematicals Abilities in Compulsory Schooling), that took shape as well as feasibility study, within an Italian context, of an IEA investigation on Mathematics. The research project, while taking over from the IEA studies their general methodological trim, created a *standardized test*, named VAMIO test: it is an entrance or leaving test for the first year of the high school (9th through 13th grades). He gathered data on a representative sample of eighth grade students.

The teaching fallout of the research: now the test is available for those schoos wishing to use it independently. In 1987 some 3.000 students reused the test; the CEDE gathered the replies of 1871 of these. From 1987-1988 to 1989-1990, 10.000 copies have been disseminated. The test was distributed on a floppy disc PC, that made it possible both to gather the CEDE data and to carry out locally the scoring and processing of the results. The software provides storing of student's answers and then immediately furnishes a profile for each student (and for all students) consistin of a series of scores and judgements. The programme provides a class profile, making a comparison with the reference population distribution. Finally, the program prints out the list of names of those students displaying some lacunas related to the various dimensions evaluated by the test.

Figure 1 shows a summary scheme of the study, of the schedule for its actuation, and of the sizes of the samples studied.

- the use of quality indicators in the field of education and training is expanding, relating both to system performance (at any level) and to individual students' achievements. Quality control can help us in better understanding, in the multiform scenary of education, the impact of socio-economic variables and the operation of system variables.
- the increasing stress on the necessity for students to reach good standards in abilities and achievements, the trend towards a generalized secondary education (at least till the age of 16), the need for new competences functional to a better insertion in society and the labour market, the necessity to rapidly change and adapt to new tasks and roles, all this has enhanced the interest in educational measurement, and monitoring procedures in education and training.

2 The Proposal of the CEDE
Building a National Test Item Bank

The *Item Banking Service*, to be established at the CEDE, aims at *prompting and coordinationg the development of measurement, regulation and control tools, and at disseminating their use, in a variety of institutional educational settings, as well as in different social contexts, where educational actions are required and particular attitudes towards self assessment are broadly shared* [1]

The Item Bank of CEDE is intended to be a *National Item Bank*, so as to play as a *National Reference Facility*. It's essentially an Information System, a Service provider, acting as interface between subjects, events, systems, anyway belonging to his operating field. CEDE's National Item Bank will operate as a *System of evaluation tools* which, as a consequence of a *continuing inquiry into the existing needs, pursues user-defined educational goals and generally follows a user-centred orientation. It forms an integrated system* for:
- *storage and management* of educational test items and related statistical data;
- *test construction* (facilities for computer assisted test construction from Item Bank on the basis of item's content characteristics and for adaptive/tailored tests);
- *analysis of results* of test administration (scoring, analyzing and reporting test results);

The project not only implies a set of centralized *general facilities* allowing storing items, processing data and retrieving items, but a *document access model* and a *document management model* (Melchiori, 1991), carried out in a reality of *networking of resources*. It should *support local banks* (schools, education and training agencies) as to usage and maintenance of test items and results analysis.

---

[1] With (for) reference to assessment and self-assessment in adult education, see Lichtner (1983 and 1988).

3 Target, Users:
To Whom?

Since the system is to be established in response to existing or potential demand for evaluation instruments, identifying the users is a preliminary task of the service itself. When the target is defined, the information needs become clearer and objectives can be stated.

As users can be globally identified as those who need utilization-oriented evaluation instruments, we prefer to define, first, the different meanings of 'evaluation for utilization' (see Fig. 1):

1) The evaluation should be support for discrete decisions. In this context, there are broadly three types of decisions having to do with program evaluation, at level both macro and educational microsystem (that is to say classroom or the individual student): decisions about program founding (procurement, change in fundin levels, renewal of funding, and initial funding requests); decisions about the nature or operation of a program (discrete actions of program staff, teaching of students, staff efficiency); decisions associated with program implementation and management (program planning, summative judgements of program impact, staff scheduling, resources arrangements).

2) Educational evaluation can be intended as project evaluation, or student evaluation; in the first sense, it aims at a better understanding of programmes and their improvement; in the second sense, it can play a formative role, or it can be functional to ...

Beyond these two meanings of utilization of the evaluation enteprise, we can speak of utilization in a third sense, related to self-assessment attitudes.

It is generally accepted that learning is internalization of external knowledge, which takes place under the control of an internal self-regulating mechanism. But after this model learning is restricted to an incremental and sequential storage of facts and definitions. On the contrary, if we take into consideration the creative and multi-source nature of learning, we have to look at it as a re-conceptualization of already internalized knowledge.

Education, therefore, cannot be passively received by the subjects. "Educational action as a possibility and as a fact is realized and sense-given by the individual, who can only act on what he has acknowledged, and that he is able to encompass in the limited field of his intentionality. For this reason the plurality of possible events, in education, has to be restricted within a dimension he can manage, and what has been so collected is to be lived by the subject as expression of a definite sense. We cannot think any longer of meaningful education as a process of mere gathering of information or as a sequential progression in acquisition of knowledge units" (Margiotta, 1989).

Following this model, evaluation should especially deal with those processes of internalization and re-conceptualization. Utilization of evaluation therefore is to be intended in a different, third sense.

3) Evaluation as self-evaluation is a sort of psychological elaboration of evaluation results which turns into an increased personal consciousness, a self-knowledge. Evaluation is "used" without necessarily inspiring decisions and producing actions.

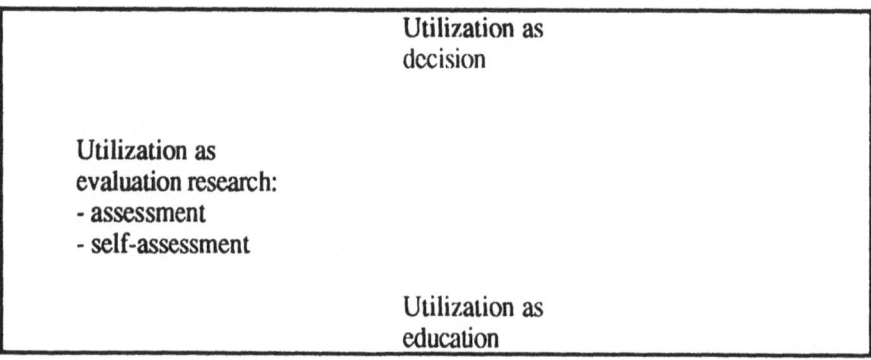

Figure 1 Evaluation utilization framework

The *"target of the Service, generally speaking, are people working in settings in any way involved in processes of enlarged social reproduction and social change, where activities of social, cultural, educational 'mediation' are at play."* (Margiotta, Master, 1992).

The framework of the virtual users arises from the cross-tabulation between the forms of evaluation utilization (fig. 1) and a variety of settings that need evaluation.

Here is a provisional list of potential users:

- monitors of public policies implying education or training actions[1];
- project leaders;
- curriculum developers;
- enterprise and labour union trainers;
- adult educators;
- researchers;
- teachers;
- teachers involved in experimental projects;
- members of examination boards.

---

[1] The most interesting case in Italy is AO2 (Azione Organica 2), a public action for the development of the South of the Italy (Mezzogiorno): see Lipari (1991).

4 Item Bank:
What's an Item Bank?

In the seventies Choppin (1985) said, an Item Bank is a collection of test items organized, classified and cataloged, like the books in a library (as an Itemtheke), from which items can be withdrawn in order to compose tests.

In the nineties such similarity reveales insufficient, as it does not highlight the essentially systematic nature of the Item Bank, which is mainly characterized as a Service System, providing information with an added value. The system produces and controles economically relevant information flows and carries out its activities according to user-oriented aims, making use of appropriate project methodologies (Melchiori, 1992).

Conclusions

Nowhere else in education are practice and theory so strictly connected as in the field of Item Banking. Progress in practice allows new theoretical developments, and advancement in theoretical models is affecting new technologies.

A particularly enhanced technology will be used, in order to optimize the potentialities of Item Banking. The management system and the interface module for user's access will be of 'object oriented' design.

References

Baldazzi, A., Indagine IEA IPS. Rielaborazione della griglia internazionale e adattamento alla realta' italiana, in "IEA IPS, Indagine sulla produzione scritta, Italia", Ricerca Educativa, V, 2-3, 1988, 1-239

Bolletta, R., Preparazione matematica in Italia al Termine della Scuola Media, Rapporto dell'indagine VAMIO (Verifica Abilita' Matematiche Istruzione dell'Obbligo), I Quaderni di Villa Falconicri, CEDE, Frascati, 1988

Bolletta, R., La Ricerca VAMIO, Ricerca Educativa, 1990, VII, 1-2, pp. 148-156

Caputo, A.M., Computers in Education: una ricerca internazionale longitudinale, Ricerca Educativa, 1989, VI, 3-4, pp. 63-114

Caputo, A.M., Ricerca CEDE sui livelli di alfabetizzazione funzionale dei corsisti delle 150 ore. I risultati dei test: l'elaborazione statistica, Scuola Democratica, XIV, 1991, 3-4, pp. 23-43

Choppin, B., Item Bank, in "The International Encyclopedia of Education", ed. by Husen & Postlethwaite, Pergamon Press, 1985, pp. 2742-2746

Choppin, B., Lessons for Psychometrics from thermometry, (posthumous 1984), in "Bruce Choppin on Measurement on Education", Evaluation in Education: An International Review Series, vol. 9, 1, 1985, pp. 9-12

Choppin, B., Principles of Item Banking (1981, 1982), in "Bruce Choppin on Measurement on Education", Evaluation in Education: An International Review Series, vol. 9, 1, 1985, pp. 87-90

Cousins, J.B. and Leithood, K.A., Current Empirical Research on Evaluation Utilization, Review of Educational Research, vol. 56, 3, 1986, pp. 331-364

Douglas, G., Latent Trait Measurement Models, in "The International Encyclopedia of Education", ed. by Husen & Postlethwaite, Pergamon Press, 1985, pp. 2911-2916.

Douglas, G., Latent Trait Measurement Models, in "The International Encyclopedia of Education", ed. by Husen & Postlethwaite, Pergamon Press, 1985, pp. 2911-2916.

Fierli, M., La Seconda Indagine IEA sull'Apprendimento delle Scienze: Prime informazioni, Ricerca Educativa, II, 2-3, 1985, pp. 89-108

Hambleton, R.K., The Changing Conception of Measurement: A Commentary, Applied Psychological Measurement, X, 4, 1986, pp. 415-421

Husen, T., IEA: An International Research Venture, Ricerca Educativa II, 1, 1985, pp. 1-18

IEA General Assembly 1988. Proceedings of the "Open Day", Ricerca Educativa, 1990, VII, 1-2, pp. 1-200

Iran-Nejad, A., Active and Dynamic Self-Regulation of Learning Processes, Review of Educational Research, vol. 60, 4, 1990, pp. 573-602

Keeves, J.P., The Improvement of Measurement for Educational Research, in "Educational Research, Methodology and Measurement. An International Handbook", ed. by John Keeves, Pergamon Press, 1988

Lichtner, M., Valutare e'necessario, EDA, N. 5-6, 1983

Lichtner, M., La Valutazione delle azioni formative, Ediesse, Roma, 1988

Lichtner, M., Competenza e incompetenza funzionale degli adulti. Una ricerca CEDE, Scuola Democratica, XIV, 3-4, 1991, pp. 5-22

Van der Linden, W.J., The Changing Conception Measurement in Education and Psychology, Applied Psychological Measurement, X, 4, 1986, pp. 325-332

Lipari, D., L'azione organica 2 come caso di politica pubblica di formazione, "Scuola Democratica", n. 1-2, 1991

Lucisano, P., (a cura di), IEA IPS, Indagine sulla produzione scritta, Italia, Ricerca Educativa, V, 2-3, 1988, 1-239

Margiotta, U., Valutazione di Sistema ed Autoanalisi di Istituto, Atti del seminario IRRSAE Emilia-Romagna, 1988, a cura di Dante Ansaloni, Bologna, 1989

Margiotta, U., Proposta die istituzione di un Master Europeo in Educational System Management, Seminario Internazionale di ricerca e proposta, CEDE, Villa Falconieri, Frascati, 1992

Melchiori, R., Apprendimento delle discipline scientifiche in Italia: Popolazione sperimentale, Ricerca Educative, 1989, VI, 2, pp. 24-69

Melchiori, R., La gestione della documentazione delle ricerche pedagogiche, Ricerca Educativa, VIII, 3, 1991, pp. 64-75

Melchiori, R., Linee Guida per l'Organizzazione e la Gestione Informativa, Ricerca Educative, IX, 1, 1992

Melchiori, R., Definizione settore Informativo, Rapporto di Organizzazione n. 3, CEDE, Frascati, 1992

Melchiori, R., Assicurazione della Qualita' al CEDE, Rapporto di Organizzazione n., CEDE, Frascati, 1992 (in progress)

Melchiori, R., Strumenti di sviluppo, Rapporto die Organizzazione n., CEDE, Frascati, 1992 (in progress)

Piracci, L., (a cura di), Proposta di adesione del CEDE al IEA Item Banking Project (1981-1985), documento interno, CEDE, 1982

Pusci, L., Progetto IEA sull'educazione Prescolastica, Ricerca Educativa, 1992, IX, 1

Van Thiel, C.C. and Zwarts, M.A., Development a Testing Service System, Applied Psychological Measurement, X, 4, 1986, pp. 391-403

Visalberghi, A., Metodologie scientifiche di ricerca in campo educativo, in"Oggetto e metodi della ricerca in campo educativo: le voci di un recente incontro", a cura di V. Telmon e G. alduzzi, CLUEB, Bologna, 1990, pp. 213-222

Visalberghi, A., Introduzione all'"Open Day" della Assemblea Generale IEA 1988, Ricerda Educativa, 1990, VII, 1-2, pp. 2-22

Analogical Evaluation Helps to Individualize Instruction

Benedetto Vertecchi

Department of Education, III University of Rome, Via Castro Pretorio 20,
I-00185 Roma, Italy

Abstract: Individualizing the educational offer is an important theme in research on both face-to-face and distance teaching methodology. The definition of the so-called "formative" function of evaluation and the development of strategies based on that funtion have taken individualization a step forward. As useful as these strategies may be, however, they leave many areas uncovered. For example, it is not enough to individualize instruction by simply compensating actively on the basis of periodic testing; this is particularly true in the case of a highly heterogeneous public. The problem may be better addressed at its roots by means of so-called "analogical" evaluation, which permits anticipating the kinds of difficulties each student will presumably encounter in attempting the goals set for a given instructional segment and thus personalizing the educational package consequently. The technique is outlined in this paper.

1 Context

The individualization of the learning proposals is an aspect of great importance in the development of the research relating both to in-presence and to distance education.

An important contribution to the development of individualization has been given both by the specification of a 'formative' function of evaluation and by the setting up of the strategies centred on this function. It is a matter of strategies which, even though they have not exhausted their validity, leave not a few intervention areas uncovered, where we cannot regard as sufficient an individualization centred essentially on the compensation activated by the training tests: in particular, these strategies show that they cannot meet the requirements of a public presenting a high degree of dispersion of its characteristics.

To face these situations an analogical evaluation is outlined here. It allows the putting forward of the judgement about the difficulties which presumably arise to prevent each student form achieving the objectives peculiar to a certain teaching route so as to individualise the entire training route with a view to meeting the needs of each student.

The definition of individualized procedures for distance education reproposes - with even greater emphasis- questions which have already been the object of attention for some time within the more general sphere of educational research.

We are not going to recall again the phases of a debate lasting for decades and which has led to formulating hypotheses and to specifying operating models of great importance; we are, rather, interested in emphasising the arrival point of such a debate and in trying to go beyond the positions which have been acquired up until now.

In fact, many signs lead us to think that the strategies of individualisation - at least due to the way in which they have been interpreted up to now - represent an aggregate which is unlikely to be expandible and that it is necessary to try new ways if we want to give new force to educational research.

It is not surprising that the potentialities, but also the limits, of the strategies of individualisation - which have been proposed over the last decades - have emerged with particular clearness in the case of distance education: indeed, we are operating within a sector of education where the rationality components in the organisation and in the management of the training processes have a crucial importance because we cannot rely on personal qualities such as sensitivity and intuition which enable teachers - in traditional school situations - to rectify the modes of intervention and to 'save' not a few educational situations.

2 Aims of individualization

The aims pursued by the strategies of individualisation are essentially the following:

a) to assure students actual equality of education opportunities supporting - to an increasingly large extent - those who show that they have difficulties in achieving the learning objectives peculiar to a certain instructional proposal;

b) to optimise the results of instruction, that is, to reach learning levels that are high and, if possible, not very scattered at the end of a certain educational procedure;

c) to favour the development in the students of affective frames of mind supporting the acceptance of the learning task trough the organisation of the training proposal which takes into account the individual needs and avoids that negative perceptions arise about the subjective adequacy to the proposed task;

d) to reconcile the collective dimension of training - which is important due to its social and psychological implications - with the need to direct to each student a learning proposal which respects his personal characteristics.

The first two aims are important above all from a social viewpoint: on the one hand, an interpretation is given which entails the right to education and on the other is stated what is expected as an effect of the increasingly great committment - at least due to the costs involved - which is necessary to support and develop the instructional systems.

The remaining two aims can instead be regarded as internal to the training activity, in the sense that they tend to specify the relationship between the student and the instructional proposal as well as to qualify the training itself as being capable of making full use of the individual differences even if within a perspective which considers the whole of the population.

3 The need for innovation in evaluation

The definition of procedures for individualisation has required that demanding lines of research should be developed.

In particular, the contribution - which could have been given to individualsation by the availability of a new evaluative theory and of tools for the verification of the students' knowledge - has turned out to be of fundamental importance.

Indeed, the shift from uniform instructional procecedure to procedures which envisage differentiated segments so as to match the individual needs, required the abandonment of rigid educational hypotheses, centred on 'models' of students which had been defined by taking into account previously - observed modal characteristics. On the contrary, it was necessary to analytically realise the range of skills which each student possesses at the time when he acceds to a certain instructional proposal, the way in which the acquisition of the skill - which constitute the specific objective of a procedure - goes forward and the learning levels reached at the end of the procedure itself.

The distinction - introduced by M. Scriven - between a formative function and a summative function of evaluation has appeared to be very opportune: if this latter accounts for the results obtained at the end of a procedure (or any how of a significant part of it), it is up to the formative evaluation to survey the information relating to the different phases of advancement 'en route'. Precisely this information enables the introduction into the educational procedure of the differentiations needed for each student to be enabled to overcome the learning difficulties encountered.

In 'in-presence' teaching the specification of a formative function of evaluation has favoured an interpretation as a compensation for individualised teaching. In other words, individualisation has consisted essentially in grafting fractions of activity - which are differentiated in relation to the needs raised through administering training tests to the students - onto a procedure whose development is common.

This interpretation has been easily extended also to distance education: it has so happened, for instance, with the organisation of the specialisation courses promoted at the University 'La Sapienza'.

4 "La Sapienza" experience

Indeed, the students are required to administer to themselves a formative-type verification test at the given deadlines foreseen by the programming of each course.

The answers wich have been furnished reveal the difficulties encountered by each student, this permitting the drafting of a text having a compensatory function.

These operations - however simple they may be from a conceptual viewpoint - require none the less a large amount of work: this explains how actual progress in individualisation has been made possible only through strong resort to technologies for automated information processing.

Indeed, to carry out the aforesaid operations swiftly and at an acceptably uniform quality-level by relyng entirely on human interventions, it would require an enormous number of operators: they ought to provide for the correction of tests and for the drafting of compensatory tests.

As each student undergoes a large number of training tests within the framework of the activities foreseen by the course he joins, the operators should write an enormous amount of texts. The automation of procedures has enabled the rationalisation of these tasks.

In the system set up at 'La Sapienza' the correction of tests and the drafting of compensatory tests are made by means of computers: a complex programme has been designed and carried out which, in addition to assuring the mentioned operations, nourishes the data file concerning the students, manages communications with them, updates the individual profile of each student, proposes continuously summary tables concerning the overall state of advancement of each course.

5 Formative issues

There is no doubt that placing evaluation from a functional point of view inside the learning process has opened up perspectives of great importance for the development of individualised teaching. The information given by formative - type tests, in fact, enables one to reorganize the teaching- learning route in order to permit each student to overcome the difficulties he has encontured as he has gone along.

In other words, the formative evalutation enables the activation of feedback every time that the performances of the students show the need for it.

It is, however, clear that the activation of the feedback will depend on the fact the student has really encountered a difficulty during his learning route and that he has given at least one incorrect performance in formative-type tests. The strategy of individualisation which is based on formative evalutation has turned out to be effective when the differentiation between the students is not excessive, but becomes rather tiring when the characteristics of the students appear to be very varied and when equally varied are the success levels that these reach during the subsequent phases of advancement in the educational route. The formative evaluation is, in fact, based on tests aiming at surveying analytically the achievement of specific learning objectives, and equally specific is the compensatory proposal which is activated by the ascertainment of incorrect performances. As the compensatory activity consists just in enabling the individual students to overcome certain learning difficulties there would be no point in supposing these difficulties can be extended to the entire route.

On the other hand, we would not understand how a sequence of specific and mutually - connected interventions such as the compensatory interventions - which are activated on the basis of the performances given in formative tests - can replace successfully the entirety of a learning route, that is an organic proposal, in which the different parts connect with one another on the basis of an accurate articulation of the content.

6 The compensatory paradigm

Briefly, the compensatory interventions can be important additions with respect to the main development of a learning plan, but it is very unlikely that they can replace it entirely or however to a considerable extent: this would happen if a student obtained a series of particularly negative results in the formative tests. Nor should the depressant effect on the motivation in the study committment - which is likely to result from the recognition of the extent of the failures - be overlooked.

While a student - who is able to give mainly - correct answers in the formative tests - feels reassured about the quality of the work - he has done and favourably accepts the supplemental work which consists in following the compensation proposals, we cannot expect that the attitude is equally fovourable if he notes generalised failures and if his subsequent work consists in recovering what he has not learned previously through supplemental activities.

These activities present themselves in segmented form, not sequential, in any case not evisaged to constitute in turn a proposal which altogether takes the place of the one where already widespread failure has been noticed.

It is true that the instructional proposal can be formulated in such a way as to contain in abundance of failures in the formative tests: this is possible by modifying the plan of instruction and by intervening on the characteristics of the formative tests.

The instructional plan can, in fact, envisage that the learning proposal utilises solutions which are more daring than the implicit components: more diffuse redundaces, more frequent exemplifications, etc.. On the other hand, the formative tests can be organised in such a way as to present a non major difficulty while also maintaining a good capability of discrimination.

But both the interventions on the instructional plan and the ones on the formative tests have as a consequence the expansion of the item which is proposed to the students: if this expansion may be in some cases (the ones which concern the students who otherwise would have met greater difficulties) functional to the achievement of more satisfactory results, in many other cases it can produce quite opposite effects, affecting negatively the affective mood of more or less considerable fractions of the students towards the learning task.

In conclusion, if the strategies of individualisation - which are based on feedback activated by formative tests - appear adequate when a not excessive dispersion of the students' characteristics is noticed, they risk not producing the intended effects or even producing opposite effects when compared to those which we would like to pursue when the students appear very differentiated in attitudes, in skills, in styles of study.

7 The individualization problem

It has turned out to be clear an actual individualisation could not consist merely in preparing compensation on behalf of those who would have revealed learning difficulties trough the formative tests.

On the contrary, it was necessary to organise the instructional route in such a way as to enable each student to receive a proposal specifically set out to meet his needs. In other words, individualisation could not be activated only owing to the performances furnished in the formative tests, but the entire learning path had to be individualised. The problem, therefore, consisted in having reliable information on the difficulties which each student would probably met in the different segments of the instructional route before this route actually began.

8 Rationale for individualization

At first, the hypothesis was formulated according to which the information needed to plan individualised routes could be obtained be means of tests proving possession of the cognitive pre-requisites in order to study the elements of statistics relating to evaluation.

This information could have been utilised to set up route segments aiming at containing the dispersion of the students' characteristics. But, upon closer examination, it has turned out to be clear that the dispersion of the characteristics not only depends on the level of possession of the prerequisites, but involves larger dimensions of the students' cognitive profile.

The previous study experiences, the acquired habits, the cultural habit which each student has developed in the course of time constitute factors of dispersion which probably cannot be compensated for merely by an initial intervention aimed at recovering cognitive abilities: instead, it is necessary to adapt the overall proposal of instruction to the individual needs, still maintaining the learning objectives which a certain route aims at achieving. It is as if to say that, in the planning of training intervention, it is not sufficient to identify the learning objectives, to set the moments of verification in the procedure and to prepare the relative compensatory activities, but it is necessary to deversify the whole proposal for it to match each student's competences, modes of approach and study style.

A new evaluative dimension is, therefore, identified for which it is necessary to develop specific tools: if we want to implement such a demanding hypothesis of individualisation it is necessary to foresee the probable behaviour of each student in relation to the learning tasks proposed in the individual segments of which the instructional route consists. Let us assume that a certain route (which corresponds to a unit of a course) is broken down into a series of segments ranging from S1 to Sn: it is necessary to foresee for each segment specific developments that are adequate to the characteristics of certain students.

We will have, therefore, S1a, S1b, etc.. up to Sna, Snb, etc..

The evaluation problem consists, therefore, in having information which enables one to differentiate the S segments.

9 A case study on statistics for evaluation

A solution has been given experimentally to this problem; it consists in proposing to the students, before the beginning of the learning route, a special test which, owing to its characteristics, has been defined as analogical: the questions of which the test consists have been formulated so as to propose operations which, even if they do not require possession of the specific competences that will be acquired gradually, can be associated in an analogical way with those which will be actually urged in the learning route.

First of all, the subjects to be developed in a unit concerning the elements of statistics for evaluation have been selected. There, then, followed the definition of the following objectives:

1a)   to catch the significance of data in the evaluative practices;
1b)   to distinguish between intuitive evaluation and objective evaluation;

1c)   to discuss the ambiguity connected with evaluation;

2a)   to classify objects and give them a name;

2b)   to establish relationships between the properties of the objects;

2c)   to identify the four basic types of scales;

3a)   to assemble objects that present analogous qualities;

3b)   to compute the frequency of these objects;

3c)   to present data in tables and graphs to make reading easy;

4a)   to carry out operations with quantitative data;

4b)   to distinguish which operations are possible and which are not according to the property of the scale;

4c)   to organize data in ordinal succession and to calculate the central tendencies (median, mean);

5a)   to interpret the meaning of mode and median;

5b)   to subdivide distribution into bands according to predetermined criteria;

5c)   to operate the quartile, decile and centile distribuitions;

6a)   to evaluate the differences between data which refer to a given measurement;

6b)   to identify the deviation between individual data and values of central tendency;

6c)   to calculate measures of dispersion (range, simple deviation, standard deviation);

7a)   to utilise the measure of the standard deviation as unit of measurement of dispersion;

7b)   to construct scales by utilising the standard deviation;

7c)   to utilise the scale of Z scores;

8a)   to examine the characteristichs of the score scales;

8b)   to establish how it is possible to create a standardised scale;

8c)   to utilise the five-point scale;

9a)   to compare the scores obtained in the same test in two different groups;

9b)   to calculate the studen's t;

9c)   to establish if the results between the two groups are or are not homogeneous between them;

10a)   to construct a model of the results expected;

10b)   to organize the empirical data in conformity with the model;

11a)   to confront the classifications obtained on the same group in two different measurements;

11b)   to calculate the measure of concomitance between the two classifications,

11c)   to assess the differences which have emerged;

12a)   to construct tables in which the data deriving from a number of measurements relative to the same group are arranged;

12b)   to verify the existence of a concomitant variation between pairs of series of data;

12c)   to evaluate the significance of the relation which has emerged;

13a)   to collect the data relating to the administration of a test;

13b)   to interpret the educational meaning of the data obtained,

13c)   to distinguish - in the interpretation - the functional placing of the test in the learning process;

14a)   to utilise a test as a 'school test';

15a)   to standardise a test;

14c) to confront the data relating to a group with the standardised data;

15a) to survey data accruing from a number of measurements and relating to the same group;

15b) to examine the differences both in diachronic an synchronic dimensions;

15c) to compute summary indices in so far as the individual results and the collective results are concerned.

10 The core features of analogical evaluation

The numbers which precede the objectives indicate the segment of the unit with which each objective is associated (the identified segments are therefore 15). The letter which follows the number marks the types of operations to which the objective corresponds:

the 'a' objectives concern operations of generalization that involve systematic acquisition of skills;

the 'b' objectives urge the shift from the general to the particular;

finally, the 'c' objectives involve the capability of applying a certain competence in a problematical context.

The distinction between the objectives substantially takes up again the one proposed by B.S. Bloom for skills having a certain complexity (capability of making generalization, making tramsformations and adaptations, solving problems).

In the analogical test a question has been made to correspond to each objective: it urges a performance which is contiguous compared with the one which is directly implied by the objective itself, but which can be given without having a specific competence. To each segment of the unit, therefore, corresponds a certain variety of behaviours on the part of the student: it is the variety of behaviours, noticed through the analogical test that permits the individualization of the instructional route (see *Flow chart of an individualized strategy based on analogical evaluation*). The conceptual formula in making an analogical test is:

$$a : p = b : x$$

where
a = contiguous field of competence
p = performance checked through an analogical test
b = specific field of competence
x = forecast of performance after a learning unit.
For each series of a, b, c objectives we have a scry of a, b, c items:

$$a' : p'_{(a, b, c)} = b' : x'$$

where
a' = contiguous skill
p' = performande checked through a sery of analogical items (a, b, c)
b' = specific skill
x' = forecast of performance after a learning segment.

Here a few examples of items (the code is the same of corresponding objective):

2a
A teacher is organizing an educational archive. Point out the group of materials that is made up by homogeneous items:
a) books, journals, videotapes, tests;
b) videotapes, slides, transparencies, films;
c) tests, practical exercices, manuals, exercice books;
d) laboratory materials, videotapes, computes, tests.

4b
In May the average temperature at 2 p.m. is 24° in Roma and 12° in Stockholm. Is it possible to say that the temperature in Rome is the double of the temperature of Stockholm?
a) Yes, because 24 is the double of 12;
b) Yes, because in Rome the mercury contained in the thermometer rises twice than in Stockholm;
c) No, because day and night do not last in Rome as in Stockholm;
d) No, because the scale of the thermometer does not begin from 0.

11c
Three cars take on the same quantity of petrol. Three drivers are then asked to drive on the same road at the same speed until the petrol has run out.
Point out the correct sentence, according to the disposable data:
a) the car going for the longest distance is a the lowest powered car;
b) the car going for the shortest distance has a motor suitable for high speed;
c) the car going for the shortest distance has a very high petrol consumption;
d) the driver of the car that went for the longest distance was particularly skilled.

If the experimentation under way is successful, it will be possible to formulate hypotheses so as to expand the resort to analogical tests to other ambits of training (for instance, the problem of equalization of the characteristics of the population acceding to a certain study iter - not only at a distance but also in presence - could be founded on new bases).

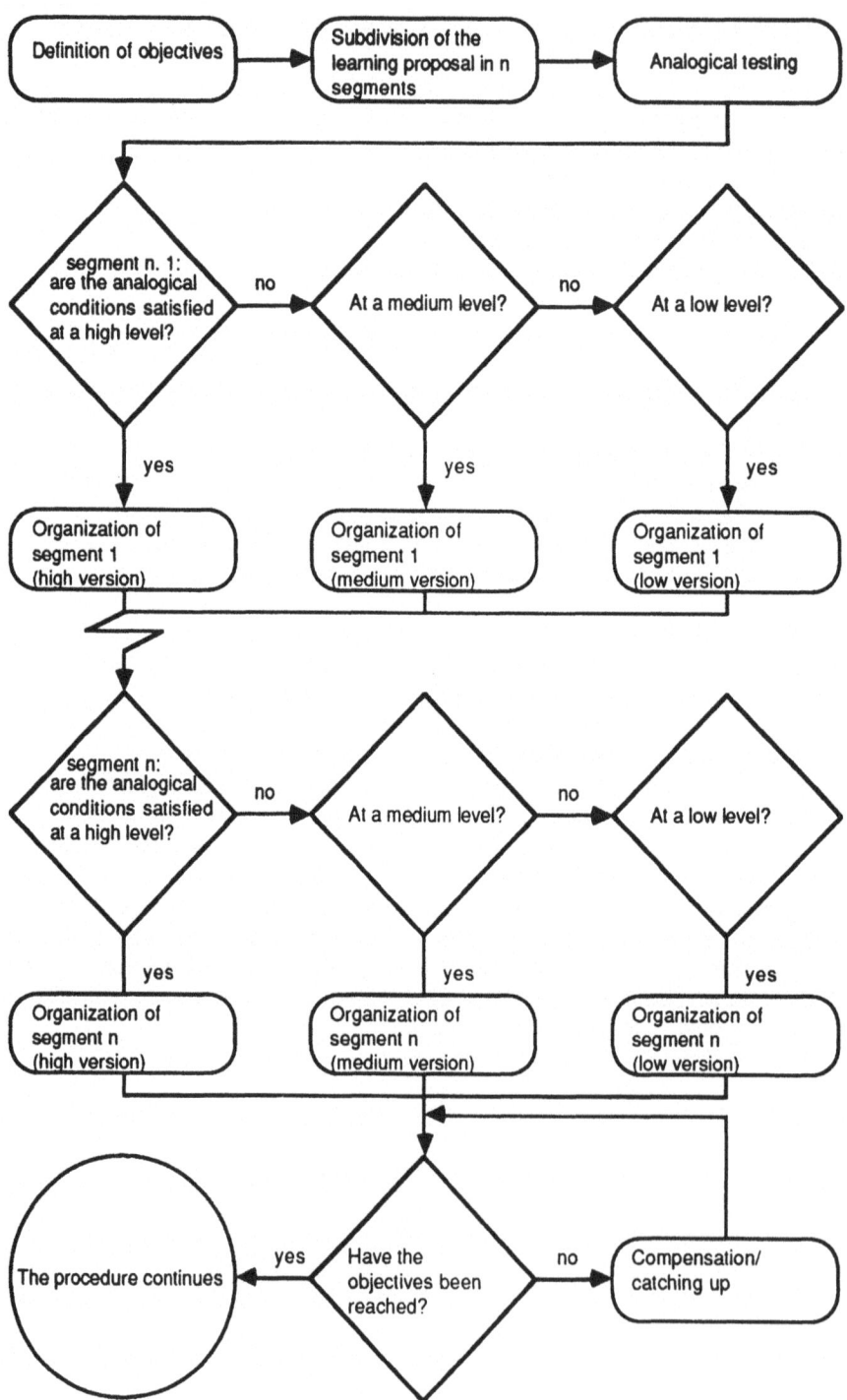

Flow chart of an individualized strategy based on analogical evaluation

11 Discussing analogical evaluation

Indeed, it is peculiar to the analogical evaluation to allow a lot of freedom in the designing of the training in training interventions. From this point of view evaluation turns out to corresponds to those requirements of flexibility and adaptability which are the crucial aspect within the most recent educational research. There exists continuity between the proposal of an analogical function of evaluation and the phases trough which docimological research has has developed up until now.

If we regard the studies carried out between the end of the twenties and the beginning of the subsequent decade within the framework of the research programme supported by the Carnegie Corporation of New York as 'a quo' term of the path of docimology, a first long phase, which continued until the fiftees, was mainly characterised by the definition of metrological aspects and by the specification of formal and technical aspects of the verification tools.

We could define this phase as systemic because the foundations of a science of evaluation have been laid during it.

The susbsequent phase, which began in the sixties and carried on through the two following decades, has seen constant involvement in overcoming the character of abstractness and separation from the training process which many evaluative practices had and therefore in placing functionally these practices within the context of the training activity.

This second phase could be defined as 'instructional'.

Now we find ourselves before the signs of a new evolution.

The fact that the training processes become complicated, the increasingly major role played by the educational technologies, the requirements of flexibility connected with the social demand for training, the push towards individualisation qualify evaluation as an activity oriented to setting out hypotheses and taking decisions.

We would like to propose defining the phase whose characters have been pointed out as 'a project' phase.

The analogical evaluation, of which a first essential presentation has been done in these pages, intends to contribute to enhancing the capability of devising training.

References

Andreani Dentici, O: *Abilità mentale e rendimento scolastico*, Firenze, La Nuova Italia, 1968.

Bloom, B. S. et Al.: *Taxonomy of Educational Objectives. The Classification of Educational Goals*, Handbook 1, *Cognitive Domain*, New York, McKay, 1956.

Bloom, B. S., J.T. Hastings, G. F. Madaus: *Handbook on Formative and Summative Evaluation of Student Learning*, New York, McGraw-Hill, 1972.

Boscolo, P. (a cura di): *Obiettivi e valutazione nel processo educativo*, con saggi di P. Boscolo, G. Cherubini, M. Cusinato, B. De Bernardi, P. Di Benedetto, G. Favaro, L. Grossele, R. Semeraro, F. Zambelli, Padova, Liviana, 1978.

Gattullo, M.:*Didattica e docimologia. Misurazione e valutazione nella scuola*, Roma, Armando, 1967.

Pontecorvo, C.: *Psicologia dell'educazione*, Teramo, EIT, 1973.

Scriven, M.: *Perspectives of Curriculum Evaluation*, AERA Monograph Series on Curriculum Evaluation, Chicago, Rand McNally, 1967.

Vertecchi, B.:*Manuale della valutazione*, Roma, Editori Riuniti, 1984.

Vertecchi, B.: *Un doppio mastery learning per l'istruzione a distanza*, "Istruzione a sistanza", 2, 1991

Vertecchi, B.: *Valutazione formativa*, Torino, Loescher, 1976.

Visalberghi, A.: *Misurazione e valutazione nel processo educativo*, Milano, Comunità, 1955.

Hypermedia: Teaching Through Assessment

Dieudonné Leclercq and Jean-Luc Gilles

Université de Liège, Service de Technologie de l'Education, Bâtiment 32,
Sart Tilman, 4000 Liège 1, Belgique
Tel. 32-41-56 20 72, Fax 32-41-56 29 44, e-mail: U017801 at BLIULG11

Abstract: This paper illustrates in which directions hypermedia could enhance educational assessment. Barriers between learning, teaching and assessment disappear. The autonomy of the learner is increased and, consequently, his/her metacognitive activity. The item bank view is progressively transformed into a learning environment that has to be explored and that helps the learner to explore his/her own learning and assessing strategies.

Keywords: Hypermedias, assessment, exploration, detection, learning styles, language learning, MCQ, confidence, note taking.

1 Introduction

It is often claimed that school assessment is artificial, i.e., does not create the real world background in which the learners' ability should be tested.

It is true that test items are too often verbal or symbolic where they should be figurative or behavioral (to use Guilford's classification, 1956, in his famous "Structure of intellect").

Accessibility and practical reasons have kept testing away from realistic setting. Nowadays, the hypermedia approach enriches the educational assessment issues in many respects. In this paper, only four of them will be considered.

1.1 Content specificity

Some assessment situations seem to be meaningful only in real settings or, at least, in a multimedia environment.

Several examples can be provided :

- The capacity to detect dangers should be evaluated only in real settings (where dangers have to be seen or heard ...).

- The capacity to understand a spoken foreign language can be assessed only by means of oral presentation.

- etc.

1.2 Learning experience specificity

It is well known that a source of the difficulty in characterising the taxonomic level (in Bloom's terms) of a question is the fact that the tester ignores what the past experience of the testee has been. In some respect, it is unfair to ask some question of somebody if it is not known whether he/she has really be exposed to the content.

When the learner has explored a content through hypermedia facilities, the selection of items can be fairer: the channel (the media) by which the content is evaluated can be made mimetic with the channel used during presentation. If not, it means that the learner is put into a situation of media transferability.

Conversely, when the testee knows in advance what the exam will be, he/she can prepare it in a more relevant way, media permitting.

1.3 Merging of learning and assessment

Up to now, a series of barriers have existed between learning and assessment whereas it is well known from formative evaluation principles that evaluation is a kernel part of the learning process itself.

Concretely, at any point of the learning process, the learner should be able to assess him/herself to check
- what his/her needs are (PRE);
- whether he/she understands (PER);
- how well he/she has memorised or transferred (POST).

This "assessment on request" contrasts with the next one, which is automatic.

1.4 The assessment of the learning process itself

In most research settings, focus is placed on the results (the outcomes) of learning whereas the actual process of learning is more rarely assessed.

Obviously, hypermedias, provided adequate reports of decisions and actions are kept, can help the teacher and the learner him/herself to understand the (sometimes largely unconscious) learning strategies sometimes referred to as "cognitive styles" (a wording we will question since it implies that permanent characteristics of the learner are the most important factors affecting learning behaviour).

* *

In the next pages, we will illustrate those four issues with specific applications and results.

2 Contents Requiring an Hypermedia Approach In Training and Assessment

2.1 The DOMUS courseware

Leclercq, Waton and Durlet (1992) have developed hypermedia courseware dedicated to the assessment of the capacity to detect dangers at home. The learner is presented the map of the house and can "walk" from one room to another either by pointing (with the mouse) at a given room on the map, or by pointing an exit door when he/she is already in a given room.

To detect a danger, the examinee has just to point it. The feedback displays reasons why this object or situation is dangerous, so that the feedback is informative and we join here the third issue: the merging of learning and assessment. In cases like this one, it seems relevant to teach through testing, i.e., put the learner in the (complex) situation, have him/her perform and then correct him/her. This should be more efficient for at least two reasons :

1. The learner can express what he/she already knows, avoiding the frustrating position of being told it as if he/she were ignorant.

2. The learner is active from the start, and evolves in successive hypothesis verification moves. Successive performances (after feedback) are likely to "naturally" exhibit progresses, improvements in ability, from simulated "experience".

2.2 The VISPA courseware

Gilles (1991) has developed hypermedia courseware that would be a component in a training program to prevent hazards and accidents in a huge Belgian drinking water production company, namely SPA The author used Interactive Video to engage the trainees in a task of detecting potential dangers

The real industrial setting has been video-recorded, from many standpoints (zooms, wide angles, inserts, close ups, travellings,) so that the learner can ask different views of the same setting as if he/she walked in it.

This shares common points with the famous Aspen experience where the MIT Architecture Machine Group has put on a videodisk the views of all the streets of a little city (Aspen, Colorado), following each in both directions, with all the possible turns in crossroads and with some "special exploration facilities" such as entering some typical houses (jail, city hall, church, school, etc.) and watching a person's interview (the sheriff, the mayor, the priest, the teacher, etc.).

Consequently, the learner can freely "explore" the town or the industrial setting through the facilities offered by the interface

3 Hypermedia to elucidate the learner's learning and assessment experiences

3.1 The DELIN approach

3.1.1 The software

The Service de Technologie de l'Education of the University of Liège has developed a software shell called DELIN (from the French word "delinearisation") to help create hypermedia-type courseware. Figure 1 shows the three important parts of the screen :
- display section (18 lines or icons);
- note taking section (4 lines);
- permanent menu section (2 lines) offering the following possibilities :
  - next screen;
  - previous screen;
  - additional information (actually an additional screen);
  - synthesis of the screen (only keywords organised in a schematic way);
  - video illustration;
  - receive a question to check my comprehension;
  - iconic (still picture) illustration;
  - introduction of notes;
  - consulting previous notes;
  - selection of a specific screen (the codes of which are provided on a paper presenting the "map" of the courseware);
  - end.

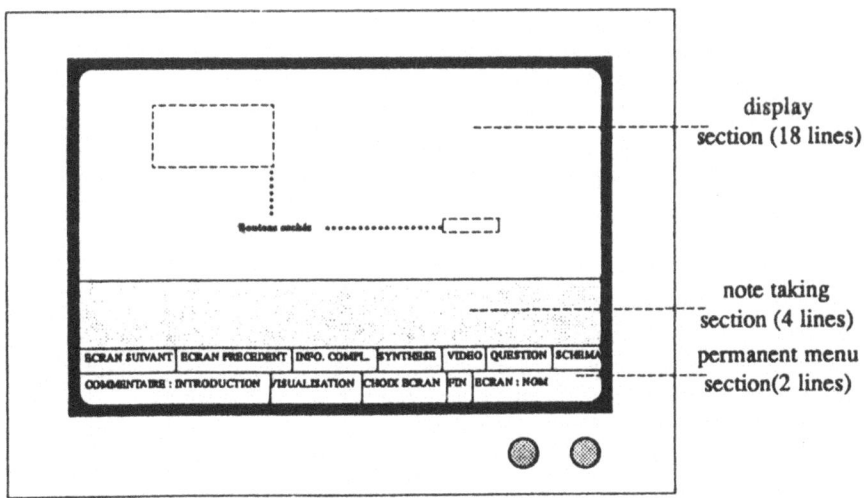

display section (18 lines)

note taking section (4 lines)

permanent menu section(2 lines)

3.1.2 The study of variations in strategies

Differential psychology trys to measure interindividual differences (the cognitive style tradition of research). Here, a more "edumetric" approach will consist in studying intraindividual variations of strategies according to variations in constraints and objectives.

Leclercq and Pierret (1989) have introduced two contents (one on Gagne's Taxonomy of Learning, or TAXGAG, and the other one on Chernobyl, TCHERNO, with a computer controlled series of video sequences).

Two groups of 9 students have been asked to learn those two contents, with two different kinds of instructions :

- *in instructions a*, learners were told that they will have to produce their own summary on what they have learned and present it orally;

- *in instructions b*, they were told that they will be evaluated through a series of MCQs.

The experimental design was as follows:

	CONTENT	
	TAXGAG	TCHERNO
Group 1 (8)	a	b
Group 2 (8)	b	a
	16	16

3.1.3 Results

Data show that in the instruction b conditions, in the two contents, students adapted
- LESS
  - linear exploration (next screen);
  - requests for additional information;
  - requests for synthesis;
- MORE
  - pointing words to know their meaning;
  - asking for being submitted to MCQs to check their comprehension.

3.1.4 Discussion

By means of a hypermedia tool like DELIN, not only can the learners self assess, but they could see the improvement of their competency after having consulted resources they like. They could even self evaluate whether they improve more when given the information in an iconic format or in a verbal format.

Confidence degrees would be a key feature in this respect since slight changes in the learner's mind would often be noticeable only through changes in confidence level. The current incapacity of measuring those switches from a state of partial knowledge to an other state of (less) partial knowledge keep educationists in the position of biologists who would not benefit from the help of a microscope!

3.2 The AUDIO-SCRIPT experiment

In collaboration with STE-ULg, the SYNAPSE team (Briol, Gilles, Gillet, Kremers and Piette) has developed a software called AUDIO-SCRIPT, that enables a teacher to prepare easily lessons to help learners to understand spoken foreign language from video recordings.

The hardware is a Sony MSX 2 with a genlocker (to superimpose the video image and the computer image on the same screen), a Umatic video recorder. The teacher has to provide a target video sequence (for instance recorded from the BBC TV programs).

The software helps the teacher
  - to delimitate video sequences of a few seconds duration (each sequence containing twenty to fourty words);
  - for each sequence, introduce (through the keyboard) the oral text so that the computer "knows" its written version;
  - introduce a dictionary of his/her own, with definition of each word;
  - inform the program of the grammatical nature of words (verb, article, noun, etc.).

This work being done, the learner can start to train in listening oral text. The screen appears with the letters of each word replaced by dots.

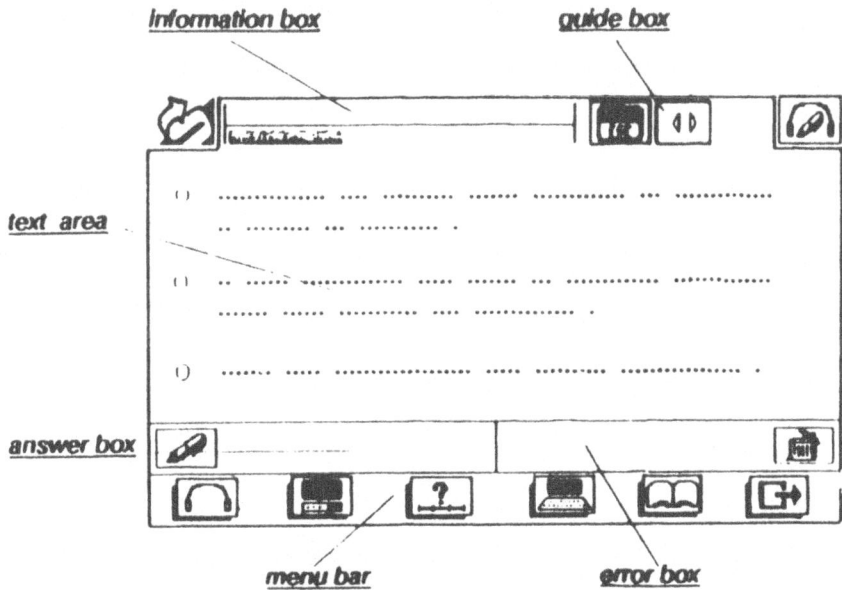

Helps offered to students are as follows :

 Hear the oral message (without vision).

 Type a word he/she has recognized. If it is one of the words really said, the typed word is placed by the program at its correct location on the screen.

 See the visual sequence in the same time as the oral one.

 Change the sequence on which to work.

 Ask for help on a specific word (to be pointed) :
  - Grammatical nature
  - First letter (in french P, from "Premier")
  - Full word (in french M, from "Mot entier").

 Ask for a "dictionary help"
  - What is the definition of this word ?
  - Does the sequence contain a word beginning with ... ?
  - Does there exist a word ending with ... ?

With the help of ISLV (1), Wauthier (1990) has implemented a series of such "lessons" on a video sequence created by Fas-Dublin, called "Developing Learning Skills". These lessons have been experimented on 4 students in education.

Wauthier (1991) has developed a graphic report to display not only the learner's strategy and performance, but also to which stimuli he/she has been exposed exactly.

The two following examples of "self-learning - self-testing sessions" display the same sequence (of 78 pronounced words, i.e. 78 words to be recognized) used by two different students (namely Pa and Sy: Figs. 1 and 2 respectively).

Each e indicates a word correctly recognized (and introduced) after an oral presentation (the student may ask to repeat oral presentations of the same sequence, that adds a column to his/her report). Each v indicates a word recognized during a sound *and visual* presentation of the TV sequence. Each P indicates a request for the first letter of the word to be provided by the computer. Each M indicates a word automatically filled in by the computer since it has been "found" by the student in a previous sequence.

Indices have been developed by Wauthier to express the learner's efficiency in hearing and understanding (with a professional interpreter being considered as the perfection level). Learners could be exposed to the successive measures of their efficiency, i.e. to their progresses and to the efficiency of their strategies, with the foreseeable impacts on motivation and on metacognition.

Nom : Fa

1 You	e			
2 learned		e		
3 these		e		
4 by		e		
5 yourself		e		
6 with		e		
7 only		e		
8 a	x			
9 little		e		
10 help		e		
11 from		e		
12 others.		e		
13 Learning	e			
14 is	e			
15 not	e			
16 just		e		
17 something		e		
18 which	e			
19 happens		P M		
20 in	e			
21 a	e			
22 classroom.	e			
23 It			e	
24 's			e	
25 something	x			
26 you	x			
27 have			e	
28 done			e	
29 every			e	
30 day			e	
31 of			x	
32 your			x	
33 life.			e	
34 At			e	
35 the			e	
36 end			e	
37 of			x	
38 this			e	
39 program,			e	
40 you	x			
41 will			e	
42 be			e	
43 able			e	
44 to			e	
45 take			e	
46 control			e	
47 of			e	
48 your			e	
49 own			e	
50 learning	x			
51 with		x		
52 confidence.			e	
53 You	x			
54 will			x	
55 have			x	
56 discovered				e
57 new			e	
58 ways			e	
59 to			x	
60 learn	e			
61 and			e	
62 you	x			
63 will			x	
64 have			x	P M
65 practised				e
66 using			x	
67 new				e
68 learning	x			
69 skills.			e	
70 This			x	
71 program				e
72 involves				
73 you	x			
74 in	x			
75 a	x			
76 number			x	e
77 of				e
78 exercises ;				

Nbre mots +	9	12	0	0		15	7		2	4	1	1	0	0	1	:53
Nbre mots -	4	3	1	0		0	2		1	0	2	0	0	0	0	:13

écouter	x	x		x	x	x x x x	=	9
voir			2		3	0
sélectionner	1				tot:	12

GRAMMAIRE			0
LEGENDE			0
PREMIERE	*	*	3
MOT ENTIER			2
MOT HAZARD		*	0
DICO ECRIRE			0
DICO SELECT.			0
DEBUT			0
FIN		tot :	4
		TOT:	16

Figure 1: Example of self-learning – self-testing session: student Pa

PREMIERE DISQUETTE troisième écran
Nom Sv

1 You	v			
2 learned	v			
3 these		p	M	
4 by	v			
5 yourself	v			
6 with			P	
7 only				M
8 a	x			
9 little		p	M	
10 help				M
11 from				M
12 others		P e		
13 Learning	e			
14 is	e			
15 not	e			
16 just	e			
17 something	e	e		
18 which	e			
19 happens	M			
20 in	e			
21 a	e			
22 classroom	e			

23 It	e		
24 s	e		
25 something	x		
26 you	x		
27 have		e	
28 done		e	
29 every	e		
30 day	e		
31 of	e	x	
32 your	e		
33 life	e		
34 At		e	
35 the		e	
36 end		e	
37 of		e	
38 this		e	
39 program,		e	
40 you			
41 will	e		
42 be	e e		
43 able			
44 to			
45 take	e		
46 control	e		
47 of	x		
48 your	x		
49 own			
50 learning	x		
51 with			
52 confidence		e	

53 You	x		
54 will	x		
55 have			
56 discovered		c	e
57 new			
58 ways		M	
59 to			
60 learn	v		
61 and			
62 you	r	x	
63 will			
64 have			
65 practised		o p	M M
66 using			
67 new			
68 learning	c		
69 skills.		e	
70 this			
71 program		v	
72 involves		M	
73 you.	t		
74 in	z		
75 a	x		
76 number			e
77 of	X	z	
78 exercises ,			

Nbre mots +   5 5   4 0 0 0 0 0 0 10 3 0 0 0 0 0 0 0   7 4   1 7 0 0 0 0 0 0 0 0 0 0 0 0 0   15
Nbre mots -   0 0   2 0 0 0 0 0 0 0 1 1 0 0 0 0 0   1 0   0 1 0 0 0 0 1 0 0 0 0 0 0 0 0   0

écouter   x x   1 x x x   x x   x   3
voir   x   1
select.onner 1 .2   2   2   tot 13

GRAMMAIRE
LEGENDE
PREMIERE   • • •   • • • • • •   • • •   0
MOT ENTIER   •   • • • • 2
MOT HAZARD   0
DICO ECRISE   0
DICO SELECT.   0
DEBUT   10   1
FIN   tot 10

Figure 2: Example of self-learning – self-testing session: student Sy

4 Merging of learning and assessment

4.1 The MASTER DIAB software

Physicians who train diabetic children (or adolescents) lack tools that will enable them to assess the children's comprehension and memory of what has been taught. It is the reason why Ackermans, Ernould and Leclercq (1985) developed a software, called MASTER (like mastery of the content) - DIAB (like Diabetes), to be used by the young learners themselves.

Children are presented the side wall of a house, with more than one hundred bricks, each brick representing a piece of knowledge, to promote the metaphor of "building" one's knowledge. The "house" has 6 floors, each corresponding to a specific content : insulin injections, urine analysis, diet, physical activity, diseases, urgencies.

The student can ask to be tested on a level of his/her choice. For each question, two parameters are specially measured :

a) *the rapidity of the answer,* since the diabetic person must react within a few seconds to avoid the hyperglycaemic coma.

This constrains has been implemented in a time scale appearing on the top of the screen for each question and a little character running along the time scale. The answer must be given in less than 40 seconds. If it is given in less than 20 seconds, the learner receives extra points.

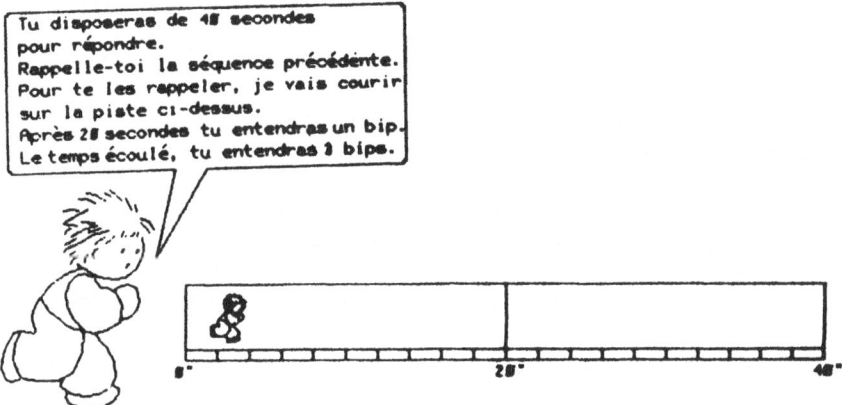

b) *the confidence in the answer* since the diabetic person has to be sure about what to do, or else ask (rapidly) for help.

A correct answer with a high confidence degree makes the corresponding brick turn blue (if given in less than 20 seconds) or green (if given between 20 and 40 seconds).

A correct answer with a low confidence degree makes the brick turn khaki, whereas an incorrect answer makes it turn yellow (if low confidence degree) or red (if high confidence degree). Those colours are not arbitrary. They are the colours to which the clinitest paper turns when dipped in urine.

Young students are motivated to "improve" their house, i.e. to fill uncoloured bricks or to make red bricks turn to a better colour (blue if possible).

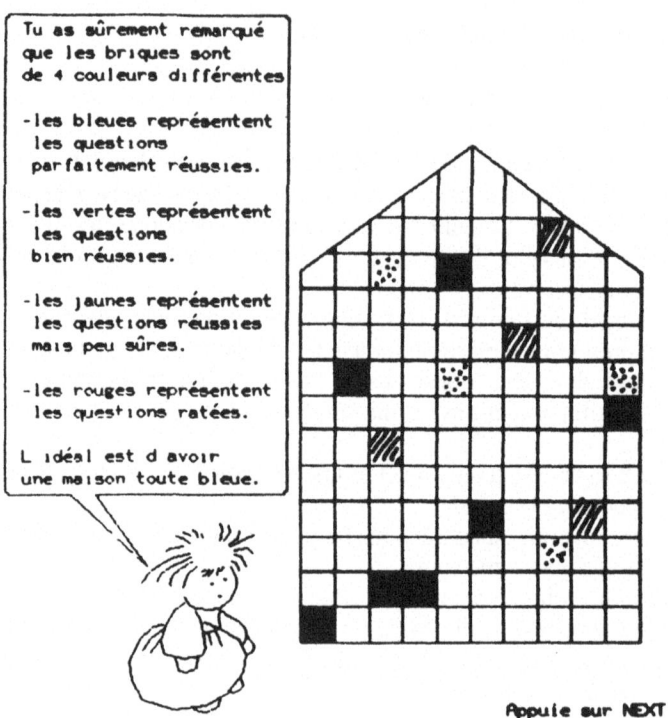

4.2 The free access to EDIP

An interesting issue in free access to item banks, i.e. in using them in a kind of hypernavigational way, is the criterion learners use to select the content on which they will be tested.

Atkinson (1964) has developed an interesting theory about the personal attractiveness of success, i.e. the pleasure different persons experience from success.

According to Atkinson's theory, there are two kinds of persons. The first ones are motivated by searching for successes; the second ones want to avoid failures. This theory predicts that the formers will choose tasks of intermediate difficulty (close to 50 %) whereas the latters will choose either very easy tasks (close to 100 %) that insure to avoid failure, or very difficult ones (close to 0 %) where failure is certain but does not elicit guilt.

Lucassen (1989) has developed an item pool, accessible by computer through a program called EDIP (Boxus and Orban, 1985) on french language at primary level. He observed that his pupils (10 to 12 years old) used to choose easy tasks, as if they wanted to avoid failure, or to experience success. This fits with Atkinson's theory because all those students had previously experienced failures and were likely to try to avoid them.

5 The assessment of the learning process itself

We will develop here only one example: a research on *note taking* behaviour *in the DELIN environment.*

The content
The DELIN "shell" has been filled (Leclercq & Boskin, 1990) with a specific content: Psychological Research on Human Visual Perception.

5.1 The possibility of visualising the strategies

The content is supported by 200 different screens, structured by a "central itinerary" and additional "lateral" branches varying in degree of depth. The figure below shows three different users' personal pathways:
(1) Continuous requests of synthesis (S) and drawings (D) = visual approach.
(2) Superficial (global) exploration to get rapidly a general view = holist approach.
(3) Systematic requests for additional data (DA) = serialist approach.

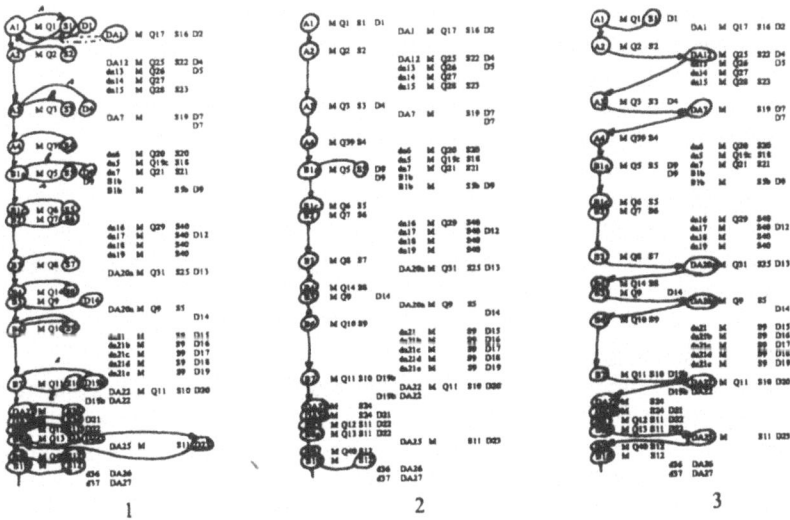

5.2 The experimental design about note taking behaviour

We wanted to study the behaviour consisting in "taking notes in order to help further consultation of the document", just as one usually does when reading reference books (inserting pieces of paper between 2 pages, folding corners, writing signs in margins, etc.) or when viewing video (noting number count and key words to help remembering sequences).

Graduate school students were asked to explore the content, using freely the possible itineraries and taking notes (on the 4 available lines) with the help of the keyboard. They were told that two weeks after, they would be asked to answer a MCQ test, their electronic notes being made available to them in a paper format.

5.3 The Test: MCQs and confidence

The test contains 15 MCQs, 8 of which been usual (the correct answer is one of the printed alternatives), the 7 others been "general implicit solution" (see Boxus, 1988), i.e. either code 6 (None is correct), 7 (They are all correct), 8 (Lack of data to decide) or 9 (An absurdity in the stem makes the whole question meaningless).

In addition, the students had to provide a confidence degree for each of their questions, on a 6 levels scale (See Leclercq, 1982 and 1988). Tariffs are computed according to decision theory so that students are interested in telling the truth (express their subjectively estimated confidence without bias). Tariffs range from -20 to +20 for each answer (one per question). Mean students scores will be expressed on a 20 points scale, averaged over the 15 questions.

A maximum of 15 screens could be consulted in the testing phase!

5.4 The results (50 students)

a) Annotated screens
Number of screens noted by students during the learning phase vary from 6 to 48, with a mean of 25 (26 by girls, 23 by boys).

Annotated screens are "central screens" (72 %), deepening (15 %), synthesis (10 %) schemata (1,5 %) and questions (1,5 %).

b) Students' opinion
The majority (47 out of 50) of the students commented that the conditions (only four lines, keyboard, severe restriction on characters, ...) made this note taking situation unusual.

c) Recognizing one's notes
Just before the test, the students were invited to retrieve *their own* notes among a series of 10 other ones (all computer printed); 47 students out of 50 were able to recognize their personal notes from the other ones, on the basis of the (declared) following cues :"my abbreviations" (15), "my writing style" (5), "the content" (5), "structure and highlighting" (5), "itinerary" (4), "my genuine spelling errors" (sic) (4).

d) Overall test achievement
Pre-test and Post-test questions were strictly identical. Mean scores (computed with confidence marking tariffs) are : 8,03 on Pre-test; 11,82 on Post-test; that is, a gain of + 3,81.

e) Facilities of groups of questions

For each kind of questions, the objective facility (percentage of correct answers) and subjective facility (average confidence degree) indices were as follows (NQ = number of questions) :

	NQ	Objective Facility			Subjective Facility		
		Pre	Post	Gain	Pre	Post	Gain
Usual	(8)	34	62	+28	47	69	+22
New Reject	(2)	22	58	+36	45	73	+28
All	(2)	52	66	+14	52	78	+26
Lack	(2)	23	30	+ 7	55	70	+15
Absurdity	(1)	52	64	+12	51	69	+18
Average	(15)	34	58	+24	49	71	+22

f) Students performances according to their characteristics

Students familiar (28) with MCQs gained more than non familiar students (22) : 4,73 instead of 2,64 in average (on a maximum of 20 points).

Students with previous knowledge of the content (13) gain less than more ignorant students (37) : 2,59 instead of 4,19 in average (on 20 points).

Students with negative attitude towards computers (11) gained less than students with neutral attitudes (39) : 2,41 instead of 5,27 in average. Students who had read a book on the content before (4) gained more than students who did not (36) : 4,72 instead of 3,78 in average.

These differences show how sensitive to conditions are the results of a testing.

g) Screen annotation and gain

Consulting annotated screens improves mean number of correct answers for 43 students out of 50 (86 %). Screens have been consulted in 35 % of cases for a correct answer (on Pre-test) and 65 % for an incorrect answer).

h) Screen consultation and lack of confidence

The relation between the confidence degree (on Pre-test) and the frequency of consulting screen at post-tests is as follows :

When confidence degree was	Rate of consulting has been
0 (low)	61 %
1	52 %
2	63 %
3	58 %
4	48 %
5 (high)	26 %

These results support Descartes' view (1636) that "doubt is the incentive of knowledge" (pp. 126-127 in the 1952 edition): the consulting behavior is explained by subjective reasons, not by the "objective" state of our knowledge.

6 Conclusions

6.1 New perspectives for research

As has been seen, students' behaviours are influenced by a series of variables. Knowing the effect of each of them helps interpreting the fundamental variables under study. DELIN (and this kind of software) appears to be a very powerful tool to observe learning strategies, be they conscious or not.

Up to now, researchers were afraid to get too subtle and too numerous data from the learner, since they would not be able to process them. Now, with interactive computer facilities, these large amount of data can be processed *on line*. Realism for instance (see Leclercq, 1993) can be computed on a continuous basis and used in the student model of an intelligent tutoring system (ITS).

6.2 Putting the driving wheel in the learner's hands

Nevertheless, an advance in the direction of the learner's autonomy with self assessment should not be counterbalanced by a backstep by having a "big-brother-like ITS" controlling information presented to the learner. It it his /her choice we want to make more conscious, informed, efficient. If a learner wants to be driven in a tutorial way, let him/her decide it him/herself, as well as coming back to the browsing, free navigating mode.

6.3 Towards new kinds of items

Since getting information (on request) and assessment can be performed by the same channel (the hypermedia), questions will look more and more like case studies. Here is its description. What is your diagnostic? Do you want additional data? In another format? A flavor of that kind of approach can be found in the TASTE methodology (Leclercq et al., 1993).

References

Ackermans, A., Leclercq, D. & Ernould, C. (1985), Un didacticiel d'évaluation des connaissances sur le diabète chez l'enfant et l'adolescent : MASTER DIAB, pp. 165-178, in Leclercq, Deghaye et Malchair (eds.), Handicaps et Technologies, Liège : Collection Education Santé, Ed. Ceres, Université de Liège

Atkinson, J. (1964), An introduction to motivation, Princeton: Van Holland

Boxus, E. (1988), Les QCM à solutions générales au service de l'évaluation à livre ouvert, in: Actes du colloque international "Formation, Evaluation, Sélection par Questionnaires Fermés", Marne-La-Vallée: ESIEE, vol. 1, 318-331

Briol, P. (1985), "Big Brother Is Helping You", ou "Quand la vidéo interactive donne à Sherlock Holmes l'allure d'un professeur d'anglais new-look", mémoire de licence en Formation des Adultes et Education permanente, Université de Liège

Brown, G. (1977), Listening to spoken English, London: Longman

Crosley, K., Green, L. (1989), Le "design" des didacticiels, Paris: A.C.L. Editions

Descartes R. (1636), Discours de la méthode pour bien conduire sa raison et chercher la vérité dans les sciences

Descartes R. (1628), Régles pour la conduite de l'esprit

Gilles, J.L. (1991), VISPA (Vidéo interactive au Service de la Prévention des Accidents du Travail à Spa Monopole). Essai de réalisation en milieu industriel d'un prototype de séquence de formation à la détection de situations de travail "accidentogènes" alliant un ordinateur et un magnétoscoope, Mémoire de licence en sciences psychopédagogiques, Université de Mons-Hainaut

Guilford, J.P. (1959), Three faces of intellect, American Psychologist, 14, 469-479

Guilford, J.P. (1967), The nature of human intelligence, New York: McGraw-Hill

Kagan, J. (1965), Impulsive and reflecting children : significance of conceptual tempo, in Krummboltz, J. (ed.), Learning and the educational process, Chicago: Rand Mc-Nally, 133-161

Leclercq, D. (1980), Computerised Tailored Testing : structured and calibrated item banks for summative and formative evaluation, European Journal of Education 15 (3), 251-260

Leclercq, D. (1983), Confidence marking, its use in testing, in Choppin and Postlethwaite (eds.), Evaluation in education, An international review series, vol. 6, 161-287, Oxford: Pergamon

Leclercq, D. (1990), Intelligent tutorial and self training system, in Proceedings of the International AI Convention, Nagoya, Nov. 1990

Leclercq, D. (1991), Hypermédias et tuteurs intelligents : vers un compromis, in G. Baron & B. de la Passardière, Actes du Colloque Hypermédias et Apprentissages, Chatenay-Malabry, Sept. 1991

Leclercq, D. (1992), Psychologie éducationnelle, STE, Université de Liège (5th edition)

Leclercq, D. and Pierret, D. (1989), A computerized open learning environment to study intrapersonal variations in learning styles: DELIN. In Estes, Heene & Leclercq, New pathways to learning through educational technology, Proceedings of the 6th International Conference on Technology and Education, Orlando, Florida, March 1989, vol. 2, 268-272

Leclercq, D. and Boskin, A. (1990), Note taking behaviour studied with the help of hypermedia, in Estes, Heene and Leclercq (eds.), New pathways to learning through educational technology, Proceedings of the 7th International Conference on Technology and Education, Brussels: March 1990, 2, 16-19

Leclercq, D. and de Brogniez, Ph. (1990), A fresh look on confidence marking, in Estes, Heene and Leclercq (eds.), Proceedings of the 7th International Conference on Technology and Education, Brussels: March 1990, 1, 646-649

Leclercq, D., Boxus, E., de Brogniez, Ph., Wuidar, H. & Lambert, F., (1993), The TASTE approach: general implicit solutions in multiple choice questions (MCQs), Open books exams and interactive testing, in Leclercq and Bruno (eds.), Item banking: Self assessment and interactive testing, NATO ASI Series F, vol. 112. Berlin: Springer- Verlag (this volume)

Pask, G. (1976), Styles and strategies of learning, British Journal of Educational Psychology 46, 128-148

Piette, S. (1985), WHY not make a better use of USE ? Essai d'amélioration du didacticiel USE (Understanding Spoken English, 1982) pour des apprenants adultes, mémoire de licence en Formation des Adultes et en Education Permanente, Université de Liège

Psotka, J., Massey, L.D. & Mutter, S.A., (eds.) (1988), Intelligent Tutoring System: lessons learned. Hillsdale, NY : Lawrence Erlbaum Associates

Riding, R. & Taylor, E. (1976), Imagery performance and prose comprehension in seven-year old children. Educational Studies 2, 21-27

Schulman, L.S. and Keisler, E.R. (1966), Learning by discovery. Chicago: Rand Mc-Nally

Wauthier, S. (1991), Développement d'une séquence de vidéo interactive pour l'entraînement à la compréhension de l'anglais oral à l'aide du logiciel AUDIOSCRIPT, Mémoire de licence en Sciences de l'Education, Université de Liège

Witkin, H.A. (1950), Individual differences in case of perception of embedded figures, Journal of personality 19, 1-15

Graphics and Verbal Items in Item Banking: Are They Really Needed?

Antonio R. Bartolomé

Departamento de Didáctica y Organización Escolar, Universidad de Barcelona, 08028 Barcelona, Spain

Abstract: This paper asks why the use of visual messages during the learning process does not correspond to a similar use during the assessment process. This paper includes a research work comparing similar items based on visual or verbal messages, within a criterion test of statistics contents. The results show significant differences between the two forms of some items, but not for the whole test. Other results are discussed.

Keywords: Evaluation, measurement, criterion referenced test, statistics, visual items.

1 Some Theoretical References

Visual or verbal messages in an item are not the same thing. The relation between image and thought has been analyzed from a Piagetian perspective (Sinclair, 1978), from a psychometric perspective (Guilford, 1977) and from an information-processing perspective (Wagner, and Sternberg, 1984). Some psychological studies from Gringer and Bandler (Patton, 1981) suggest that different persons have developed different verbal, tactile, auditive and visual systems for their representation and interpretation of the world. M.Q. Patton concludes that there is not a unique and effective modality of communication and experience relevant for everyone.

The image does not always help to solve an item. Denis (1984) picks up some examples from the work of Wohlwill and of Winner and Kronberg. In our study, the introduction of visual elements is referred to an effective codification of item essential information. According to Macdonald-Ross (1978) we do not consider the use of icons for motivation as it is the case in cartoons.

2 Research Planning

The basic tool was a multiple choice criterion-reference test (Popham, 1975, 1978). For each specification we prepared two similar contents: I and II. For each content we developed one item in two forms: iconic and verbal; an example follows.

Specification:

Referring to sampling distribution of
  . Proportion of one category from two
  . One qualitative variable with several categories
  . Mean from a quantitative variable
to recognize
  . model of probability distribution.

Content I, Visual Form
Select the model for the sampling distribution of one category:

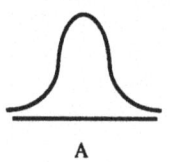 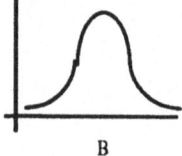 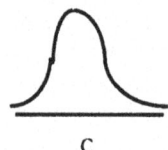 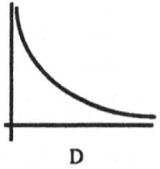

| A | B | C | D |

Content I, Verbal Form
Select the model for the sampling distribution of one category:
  A) Normal Distribution
  B) Skewed Normal Distribution
  C) t distribution with 1 degree of freedom
  D) c^2 distribution

Content II, Visual Form
Select the model for the sampling distribution of the mean:

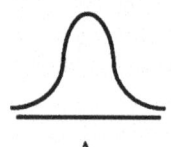 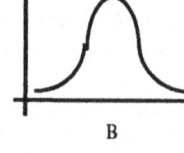 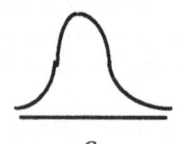 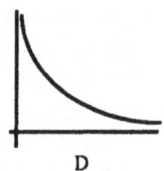

| A | B | C | D |

Content II, Verbal Form
Select the model for the sampling distribution of the mean:
  A) Normal Distribution
  B) Skewed Normal Distribution
  C) t distribution with 1 degree of freedom
  D) c^2 distribution

Two questionnaires were prepared from this item bank. The randomization included some restrictions: for each item, the two forms were assigned to one or other questionnaire at random, and the other content was automatically assigned to the opposite form. So:
. if specification 13 was randomly assigned to questionnaire A
  with content I and VISUAL form (13 - I - VISUAL to A)
  then...
  . 13 - I - VERBAL is assigned to questionnaire B
  . 13 - II - VISUAL is assigned to questionnaire B
  . 13 - II - VERBAL is assigned to questionnaire A.

The sample was composed by 46 students of Statistics for Educational Sciences. Each student answered one questionnaire, A or B. The tests were randomly distributed according to a systematic plan over a spacial distribution. The students had to solve also a classical test based on problems solving.

3 Test Design and Development

The test aims to assess a specific domain: the use of inferential techniques on Normal, t-Student and Xi-square. The framework for preparing the tool was as follows:

1. Content analysis of 4 tests for similar contents from other professors and years
2. List with 16 skills from these tests
3. Comparison of skills with the objectives of the course
4. List with 13 basic skills
5. Evaluation of that list by 4 professors of Statistics
6. Definitive SPECIFICATIONS
7. First redaction of items
8. Contents I and II for each item
9. Forms VISUAL and VERBAL for each content
10. Randomly items assignation
11. Definitive QUESTIONNAIRES A and B

The reliability was studied in a similar way to the one proposed by Popham (1978, pg. 206). The criterion was established in 70% of success for expert. We obtained an estimated reliability coefficient between 0.551 and 0.601 (a = 0.05). The two models of questionnaire were compared. According to the no-normality of test, we used a non-parametric tool: U of Mann-Whitney. We obtained $z=0.76$ (p = 0.45).

The content validity was obtained during the process of development of specifications. According to Mehrems (1973) was this validity the most important for the criterion referenced measure and it was carefully considered. From a functional perspective we studied the correlation between the test and a classical test of problems. We obtained a value of Phi=0.323, significant at 0.05 level.

A last observation: In contrast with the norm-referenced tests, there is a lack of efficient and safe tools for working with criterion reference test.

4 Analysis

We find differences visual vs. verbal (content I vs II) for each individual: we find only significant correlation coefficients in 3 out of 46 subjects. We could have concluded that there were differences in the way students answered visual and verbal forms. However, it is necessary to make two observations.

First, according to Popham, in a criterion-referenced test, the weight passes from the individual items to the whole test. So, it is not so interesting to check whether when one student gets right the visual form, he also does the verbal form, but whether we can consider as experts, according to the answers to verbal items, the same students we could consider as experts according to their answers to the items under visual form. In this way, the Phi coefficient for the whole test between visual and verbal items was 0.5167, significant with a = 0.05. The 77% of the students got the same measure from their visual items or from their verbal items.

Secondly, for the same specification for each subject, the visual and verbal items differ not only by their form, but by their contents, I and II. So, we analyzed the difficulty value of items separately for specifications, contents and forms. From this information we compared forms for the same content and contents for the same form. We found 4 significant difference between forms and 9 between contents (a = 0.05). The differences between contents grew under verbal forms. These items were analyzed, and we have included some interesting observations in the results.

The success ratio was 0.61 for visual items and 0.63 for verbal items. We did not find significant differences between forms in the way success was distributed over the specifications.

5 Results

The main result is that for the whole test, the different forms do not affect the assessment of students.

In some items we have found significant differences. These suggest that specifications could include some information about the need of graphics.

The observed differences between contents that were designed as similar suggest that the use of items banks and the selection at random is the best way for preparing alternate versions of the same test. It seemed that visual forms reduced the difference between contents. From this, we can question whether verbal forms have more discriminating power and whether this discrimination is relevant.

References

Denis, M. (1984) Las imágenes mentales. Madrid: Siglo XXI.

Guilford, J.P. (1977) Way beyond the I.Q. Buffalo: The Creative Education Foundation Inc.

Haertel, E. (1985) Construct Validity and Criterion-Referenced Testing. In Review of Educational Research, vol. 55, 1, 23-46.

MacDonald-Ross, M. (1978) Graphics in texts. In Review of Research in Education 5,49-85. Itasca (Ill): Peacock Publishers.

Mehrens, W.A. (1973) Measurement and Evaluation in Education and Psychology. New York: Holt, Rinehart and Winston.

Patton, M.Q. (1981) Creative Evaluation. London: Sage Publ.

Popham, W.J. (1975) Educational Evaluation. Englewood Cliffs (NJ): Prentice-Hall Inc.

Popham, W.J. (1978) Criterion Referenced Measurement. Englewood Cliffs (NJ): Prentice-Hall Inc.

Sinclair de Zwart, H. (1978). Adquisición del lenguaje y desarrollo de la mente. Barcelona: Oikos Tau.

Wagner, R.K. and Sternberg, R.J. (1984) Alternative Conceptions of Intelligence and their implications for Education. In Review of Educational Research, 54, 2.

Adaptive Testing : Contribution of the SHIVA Model [1]

Romain Zeiliger

Centre National de la Recherche Scientifique, Laboratoire IRPEACS, Institut de recherche pluridisciplinaire sur les environnements d'apprentissage et de communication de savoirs, 93 Chemin des Mouilles BP 167 - 69131 Ecully Cedex-France,   Tel (+33) 72.29.30.09, Fax : (+33) 78.33.33.70

Abstract: SHIVA is a prototype courseware development environment developed within the framework of EEC Delta project. Using both an embedded teaching strategy and a provided conceptual network of the domain, the SHIVA system enables an author to produce an adaptive sequence of multi-media teaching and testing units. This paper describes the SHIVA model and the components of the Microsoft Windows inplementation, and the implied learning, teaching and authoring processes. It then emphasizes the mechanism of the assessment units and their link with the student model and the teaching strategy. It ends with a discussion on the SHIVA characteristics which could contribute to the design of an adaptive testing system.

Keywords: Multimedia, authoring, ITS, adaptive systems, student model, curriculum design.

1 Introduction

SHIVA is a multimedia *intelligent* authoring environment which has been developed as part of the DELTA project D1010, *Advanced Authoring Tools*. The implementation that we describe hereby was made at IRPEACS laboratory and runs on Multimedia Windows 3.1 (from Microsoft).

SHIVA results from an attempt to 'bridge the gap' beetween well understood CAL educational technologies and research tools in ITS (*Intelligent Tutoring Systems*) (Elsom-Cook 1989) by implementing the ECAL system (the Open University, GB) as a mechanism for making pedagogical decisions within existing multimedia CAL (*Computer Assisted Learning*) materials.

---

[1] This paper draws on work carried out for the AAT project (DELTA 1010). The consortium for this project consists of CNRS-IRPEACS (Lyon,France), Open University (Milton Keynes, UK), Apigraph (Lyon,France), SEL (Pforzheim, Germany) and DATAMAT (Rome, Italy).

SHIVA cannot be considered as an ITS because it is not capable of generating nor modifying its teaching materials. It only performs a sequencing of those materials. It fits with the definition of "one-on-one tutoring systems" according to G. McCalla (McCalla 1992) : SHIVA has access to a representation of the domain knowledge (a semantic network) ; it has a tutoring strategy that takes advantage of this knowledge to produce a curriculum of instructional activities.

Testing systems commonly use a bank of questions to assess the knowledge of students regarding a particular domain : a "test" about a given topic comes as a sequence of questions chosen from the bank. Curriculum is essentially a control path through the items of the bank and must be explicitly predicted when designing the test. It causes a certain rigidity of the system and leads to a kind of curriculum which is not particularly individualized nor adaptable. It is far from what we can observe in face-to-face oral examination.

The matter of this paper is to examine the mechanism that support adaptability in SHIVA and to discuss whether it could contribute to the design of testing systems, both in summative and formative student evaluation.

2 SHIVA description

2.1 SHIVA components

A standard accepted architecture for an ITS includes four major modules : a domain expert, an instructional expert, a student model, a teaching interface (Psotka,Massey,Mutter 1988).Despite earlier efforts to keep these major modules separate, experience has shown that their embedded knowledge is highly interrelated.To perform an effective teaching, the system requires to link knowledge available from different modules. From a software engineering perspective, the major difficulty in designing the system is to define the interrelationships that will enable the four modules to cooperate.

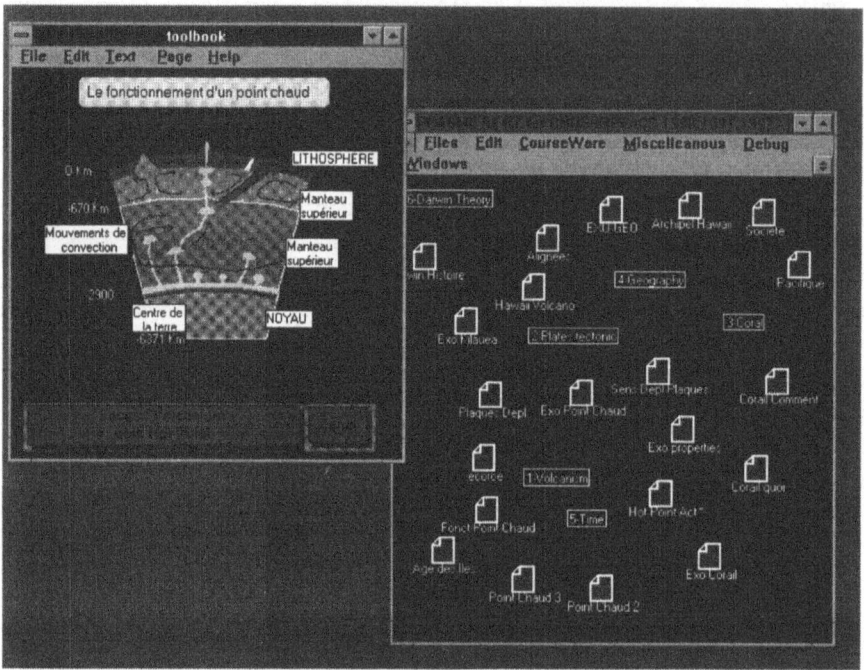

Fig.1 SHIVA authoring interface : the ILO-ULM network (right) indicates the current focus ("Volcanism") and the current ULM ("Fonct Point Chaud") while this current ULM is displayed to the student through ToolBook (left).

A staightforward difference beetween ITS and CAL systems concerns the separation of domain knowledge from the independant set of processes which operate upon it to generate teaching materials and control their sequencing in a pedagogical interaction. Within the SHIVA project, we made the assumption that this separation of domain and teaching knowledge in the system, translates simply into a separation of domain presentation and teaching knowledge within the authoring process : the author controls the former, and the system the latter. We assumed also that the domain presentation can be split into small "chunks" of teaching called "Units of Learning Material " (ULM's).

For a specific domain, the SHIVA environment components come as follows :
- a set of ULM's presenting the domain : they are small self-contained programs, involving multimedia presentations and interactions (text, graphics, sound, video sequences). They fall into two classes : The *presentation ULM's* do not involve any pedagogically important interaction with the student (ex : re-play sound, next,...), or even no interaction at all. The *Diagnostic ULM's* offer interactive testing of the student knowledge. Each ULM defines its own teaching interface, which means that consistency has to be cared for.
- a domain dependant conceptual network of the pedagogical objectives to be achieved in the teaching process : within the SHIVA model we call them :

ILO (for Intended Learning Outcome). Each ILO is linked to one or more ULM's (fig 2). Those ULM's which are linked to a given ILO are assumed to provide the teaching "chunks" that will enable the student to learn the correspondant knowledge. A possible ILO hierarchy is supported in the form of prerequisit links.

This model is partly derived from curriculum design techniques and is based upon that of Posner and Rudnitsky. (Posner,Rudnitsky 1985) : in the original P & R model the author identifies key concepts of the material to be presented and then clusters these concepts into groups based on how closely related they are ; in SHIVA, the clustering is computed automatically from the network provided by the author : the system infers the relatedness of a given ILO against the other ILOs and computes the ILO "generality" array.

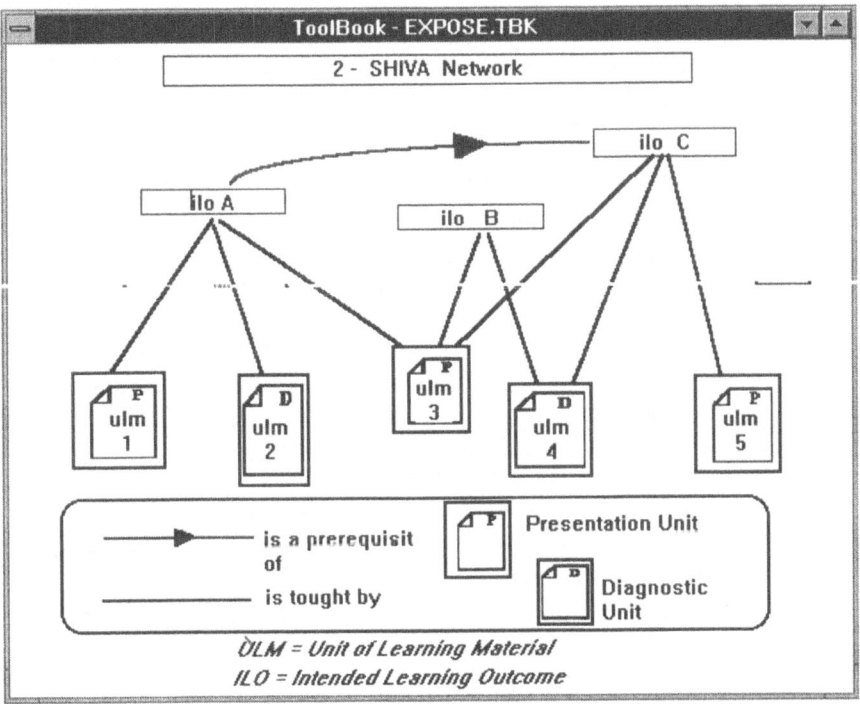

Fig 2. ILO-ULM network components : topic "ILO A" is tought by presenting material of units "Ulm1" and "Ulm2"; the student knowledge about "ILO A" is assessed with unit "Ulm2". Topic "ILO A" is a prerequisit for topic "ILO C".

Compare with the network in Fig 1 : the topic "Volcanism" is tought by 6 presentation units: "Hawaii volcano, Ecorce, Fonct Point Chaud, Point Chaud 3, Point Chaud 2, Hot Point Act"; it has 2 assessment units : "Exo properties, Exo point chaud".

- a student model using subset overlay modelling technique, mapped on the conceptual network (P.Fung, M.Elsom-Cook 1990). Based on pre-recorded learning curves, it keeps track of the student interactions and maintains a numerical model of the student knowledge.
- a generic instructional expert uses the conceptual network, the student model and an embedded dialogue model, through simple rules, to determine an adaptive sequence of teaching. It makes the decisions about the changes of focus and chooses the appropriate teaching units among those linked to the given focus. Its pedagogical embedded strategy is to *minimize the newness* of the teaching material supplied at every new step of the teaching sequence. The newness of a possible Focus or of a possible ULM is evaluated by comparing the student model with data derived from the conceptual network. The "minimising newness" rule is used to minimize the focus shifts during the teaching dialogue and thus to provide a conceptual continuity.

2.2 Learning process

Within a SHIVA produced courseware, the student learns through interaction with a sequence of multimedia teaching units. The system provides plain guidance : it does not allow the student to browse through the learning material..However the learning sequence is tailored to each individual student. The student receives information about the on-going learning process, such as the current focus and the unit name, but he is not informed about the state of the student model. Each ULM is provided to him only once. He is not allowed to review the already presented material and so he is not allowed to exercise again with the testing units he has failed in. Some of these real limitations of the system are not inherent to the model and could be broken so as to improve the learning process.

2.3 Teaching process

The sytem chooses, among the ILO of the network, a first one to start with. This is the first focus. This choice is based on an evaluation of the generality of this focus computed from the network. The first focus cannot have any prerequisit. Then, among the ULM participating to the teaching of this focus (those linked to the choosen ILO in the network) the system chooses one to start with and provide it to the student. As each ULM is self-contained, the system does not make any decision while the student is interacting with the current one. As soon as the current ULM is completed, the system updates the student model. This update process takes into account the position of the completed ULM in the network, its belonging class (presentation or diagnostic), the way the student has interacted with it, and the current position of the student on the correspondant learning curve.

The next decision will be to proceed with the teaching of the current focus by providing a new ULM to the student. The newness criterion then applies. The teaching of the current focus proceeds until all the *presentation ULM's* have been provided to the student.

Then comes the assessment : the assessment units (*Diagnostic ULM's*) which are linked to the current focus, may be linked to some other ILO's. The system makes the decision on whether the student knows enough of the relevant topics (ILO's) to be tested. When no assessment can be applied, the system has to choose a new focus on the base of minimizing the newness criterion and proceeds in the teaching of this new focus. The system will eventually come back to the assessment of the not-completed old focuses. Since the knowledge of the student is mostly evaluated in the assessment units (in reference to the learning curves) the result of the assessment determines most of the focus changes. This mechanism makes the teaching-and-assessment sequence an adaptive one.

2.4 Authoring process

With the improvement and standardisation of Multimedia, a lot of high quality software tools are available from the market, which enable an author to built multimedia CAL material. As those tools are not dedicated to teaching, they lack some important pedagogical features but recent software developments, such as the client-server architecture, contribute to turn them into well opened sytems :
they can cooperate with some specifically developped software modules dedicated to the teaching, learning and authoring processes.

This new implementation of SHIVA is an hybrid system in which MultiMedia CAL is achieved by on-the-shelf software tools, while the 'ITS' part is issued from research developments. Visual authoring is achieved through graphical human-machine interface.

2.4.1 Authoring the ULMs

This task was originally supported by the ORGUE multimedia tools for DOS. When working with Microsoft Windows the author uses now TOOLBOOK (from Asymetrix) to build the Learning Units. Digital sound, video sequences, graphics and text can be used to present the domain ; interaction and interactive testing is through *direct manipulation* and can be built using script programming. The OpenScript language capitalizes on Object Oriented Programming and allows a high degree of learning material re-usability. The ULMs come as Toolbook's *books*. However, the navigation among the different pages is managed by SHIVA: in fact, any *next* button relies on SHIVA decision mechanism. When building a Diagnostic unit, the classes of the possible answers and the associated mis-understood ILOs are defined using a set of new OpenScript statements.

2.4.2 Authoring the network

An interactive graphical network editor is provided : it follows the Microsoft Windows Multiple Document Interface scheme and provides different views of the constructed network such as the prerequisit graph and the implicit relations graph. Construction of the graph is through direct manipulation. A simplified conceptual network representation can be built in order to display it - via TOOLBOOK - to the student.

2.4.3 Authoring the teaching sequence

Given a set of ULMs and given a Conceptual network, the SHIVA system determines a finite set of possible teaching sequences (teaching sequence varies with the student answers). A teaching sequence simulator (thus simulating the student answers) allows the author to forecast the teaching sequence and to verify that the generated set of sequences is appropriated to the intended teaching. To alter the teaching sequence, one has to alter the network. This is an incremental task leading to a validated network.

3 The SHIVA assessment mechanism

Within the SHIVA model every *presentation ULM* as well as every *diagnostic ULM* is linked to *one or more* of the ILOs.They are the ULM's associated ILOs. As soon as the student finishes to deal with the current ULM, the system updates the student model. When the current ULM is a *presentation* one, the system increases the confidence in the student's understanding for each associated ILO.

When the current ULM is a diagnostic one, the process is as follow :

- if the student answered correctly, the system confidence in the student's understanding is increased for each associated ILO, but slightly more than in the case of a presentation ULM.

- if the student answered incorrectly, the system confidence in those particular ILOs associated with this particular answer is decreased, while the system confidence in the remaining associated ILOs - if any - is increased slightly. In fact, any incorrect answer must fall into a pre-declared answer-class, which has been associated (during the authoring process) with a subset of the ULM associated ILOs. Those are the ones the student failed to understand.

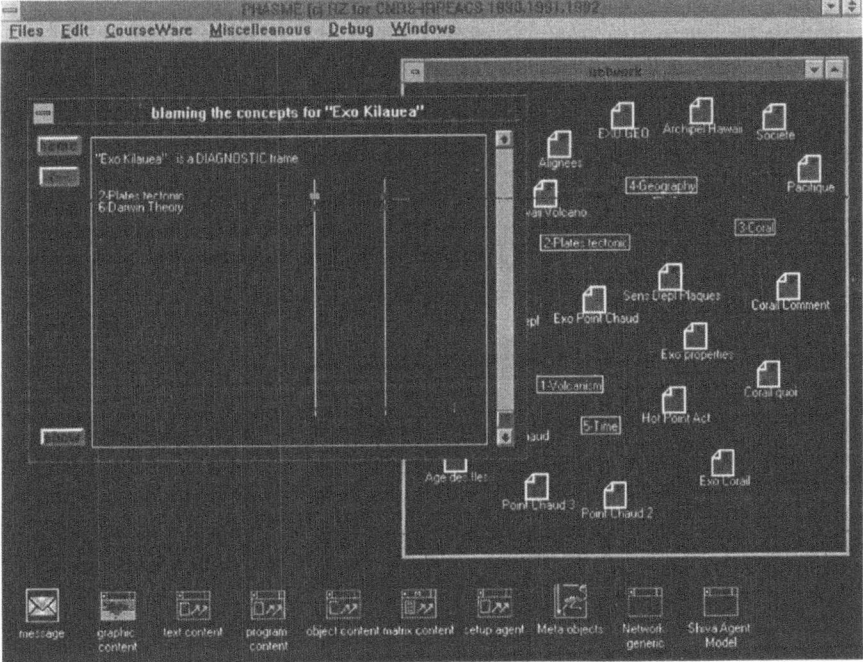

Fig 3. Defining the answer classes in a Diagnostic unit : the current unit ("Exo Kilauea") ask to the student a question connected to 2 topics: "Plates Tectonic" and "Darwin Theory" (see the network in fig 1). Two answer classes have been defined (figured by 2 vertical lines) : the first class defines the misunderstanding of "Plates tectonic" as being responsible for the answers falling in this class, the second class correspond to answers of student misunderstanding both involved topics.

Every answer class is defined through its properties : those properties include a range of equivalent answers and may include a list of range of answer characteristics such as the elapsed time-to-answer. Subjective probability distribution, when available, could be taken into account as one of the properties characterising a given answer class. A tariff could even apply when updating the student model. The important thing in the SHIVA model about answer classes is that they can be considered as associations beetween a set of answer properties and a class of student misunderstanding defined relatively to the current involved pedagogical objectives (ILOs). This association, although quite "poor" in the ITS perspective, allows to simulate the teaching sequence.

The increments and decrements in student confidence are computed from pre-recorded learning curves : when starting to learn a new topic the student progress is assumed to be fast ; progress is slowing down during the last steps that lead to full understanding of the topic.

4 Some identified limitations of the SHIVA model

As it has been said, SHIVA is only a prototype system. Here are some of the issues raised during the SHIVA evaluation task that was part of the DELTA-AAT project (Delta D1010 1991): difficulty for authors to make a clear separation beetween teaching materials and control structure ; difficulty to produce a domain knowledge representation suitable for educational purpose ; difficulty to define the optimum ULM size (from a single screen to a whole interactive sequence) :the ULM size has consequences on the accuracy of the student model and determines the extent to which the system controls the teaching sequence ; difficulty to partition the knowledge of a given domain and thus to identify ULMs ; criticism of the rules driving the tutoring strategy and their applicability to a given domain ; necessity to stand above a threshold in term of network components (ILOs and ULMs) allowing the system to work ; lack of remedial units.

5 Toward an adaptive testing system

What we call here an adaptive testing system, would be a system selecting from an item bank an adaptive test sequence to serve a specialised purpose. An adaptive questionning sequence provides the student with some questions which depend on her previous answers (for example some additional questions can be issued when the system condifence in the student understanding of a topic is not high enough). We suppose that the system does not require that the author predicts *explicitly* the sequencing at design time, but rather that the author provides some *knowledge* allowing the system to make "on the spot" decision about the next question to select from the bank. This kind of questionning sequence would come closer to what is face-to-face oral examination (Boxus 1992).

It is clear enough that SHIVA is not a testing system : it is a teaching system in which the teaching process is partly based on an assessment process. It does support individualised tailored teaching, and obviously some individualised testing to achieve it.As the demarcation beetween instructional and testing systems becomes increasingly blurred (Leclercq 1985), it could be worth reviewing some of the SHIVA key features which fit well with computer-managed Testing.

* Shiva diagnostic ULMs are close to Interactive Testing Units. However they lack some important properties such as item difficulty, format, answering instructions , which allow a test classification.
* The testing material in SHIVA is multimedia. The student can interact through mouse designation or direct manipulation as well as multiple-choice questions.

The answer classification mechanism fits with graphical as well as textual interaction.

* Any style of multiple-choice question can be implemented in the ULMs ; then answer classes properties have to be defined to match the type of the expected answers : for exemple one can define a SHIVA answer class based on a range of numerical values ; one can assign a SHIVA answer class to a combination of answers. An interactive elicitation technique for subjective probability distributions could be easilly included in some of SHIVA diagnostic units. A graphically oriented interactive elicitation, such as ELI (Van Lenthe 1992) would fit well with the SHIVA interaction style. When defining the answer classes subjective probability distribution profiles should be accepted as characterising properties.

* Shiva relies on a bank of self-contained items (the ULMs).

* Those items are clearly linked to objectives (ILOs) : the conceptual network relates every testing Unit to one or more ILO. However there is no weighting of the relations.

* The pre-requisit links establish a hierarchy among the objectives.

On the other hand, turning SHIVA into a testing system raises important issues :

* When the aim of the system is to test and not to teach (so there is no teaching ULMs) what is the initial state of the *student model* ? Do we credit the student with an initial knowledge, the role of the sytem being to assess it ? Do we infer the student knowledge from his successive answers, but then don't we have to define an initial questionning sequence, and with which strategy ? Is this student model, based on a theory of learning, still relevant to testing purposes. Is the SHIVA student model a kind of test score ?

* SHIVA has an embedded *teaching* strategy which is not obviously a *testing* strategy. This strategy, based on some testing results, allows to decide what to teach next but of course not what to test next. It seems that a testing strategy should have to take into account some specific properties of the testing units (which do not exist in SHIVA) as well as some important dimensions of the student answers such as the realism level. Supposing that subjective probability distributions is elicited within SHIVA diagnostic units, then a student realism assessment could be computed and made act upon the tutoring strategy through specific rules.

Conclusion

An important issue of SHIVA has been to define the framework of a learning environment providing such enhancements as multimedia and adaptability which represent a step beyond traditional CAL. Some key components within this frame work are not appropriate to testing and they must be fully re-designed in order to match this new goal. However we think that this redesign process could well take place within the existent framework and efficiently capitalize on it. The future developments of SHIVA include a graphical representation of its main components (on the base of agents model) thus allowing an author to modify the original model: this new feature could well be used to derive from SHIVA some prototype adaptive testing system for research purposes.

References

Baker, M. (1989) Modelling the authoring process in SHIVA, Discussion Document Delta D1010, 1-31. Ecully: CNRS-IRPEACS 93 Ch. des Mouilles PB 167, F-69131 Ecully Cedex France.

Baker, M., Bessière, C.(1990) Intelligent multimedia authoring: advancing towards users in a European context. In: Learning technology in the European Communities. Proceedings of the Delta Conference on research and development, The Hague 1990, pp. 653-659. Dordrecht: Kluwer Academic Publishers.

Boxus, E. (1992) Check: une banque de questions interactive, Colloque ESIEE: Les questionnaires automatisables, Marne-La-Vallée. Université de Liège, STE, 5 Blbd. du Rectorat, Start Tilman B-4000 Liège 1, Belgique.

Delta project D1010 (1991) Final report: advanced authoring tools, pp. 1-38, Ecully: CNRS-IRPEACS 93 Ch. des Mouilles PB 167, F-69131 Ecully Cedex France.

Elsom-Cook, M. (1990) The Ecal teaching engine: pragmatic AI for education. In: Learning technology in the European Communities. Proceedings of the Delta Conference on research and development, The Hague 1990, pp. 329-340. Kluwer Academic Publishers.

Fink, P.K. (1988) The role of domain knowledge in the design of an intelligent tutoring system. In: Intelligent tutoring systems: lessons learned. pp. 195-217. London: Lawrence Erlbaum Associates Publishers.

Fung, P., Elsom-Cook, M. (1990) Student modelling toolkits. In: Learning technology in the European Communities. Proceedings of the Delta Conference on research and development, The Hague 1990, pp. 383-394. Kluwer Academic Publishers.

Leclercq, D.A. (1985) Computer-managed testing. In The International Encyclopedia of Education, pp. 943-944. Oxford: Pergamon.

Leclercq, D.A. (1990) L'apport des nouvelles technologies a la formation, Symposium: La psychologie du travail et les nouvelles technologies, Liège 17-18 Mai 1990, pp. 1-15. Université de Liège, STE, 5 Blbd. du Rectorat, Start Tilman B-4000 Liège 1, Belgique.

Léonhardt, J.L. (1991) Multimedia authoring: some evaluation. In: D.L.T. news issue 8, July 91, 8-9. Paris: Les Editions du Logiciel d'Enseignement, 8 rue Duguay-Trouin, F-75006 Paris, France.

McCalla, G.I. (1992) The search for adapatbility, flexibility, and individualization: approaches to curriculum in intelligent tutoring systems. In: Adaptive learning environments: foundations and frontiers, NATO ASI Series, Vol. 85, pp. 91-121. Berlin: Springer-Verlag.

Posner, G.J., Rudnitsky, A.N. (1985) Curriculum design. New York: Longman.

Psotka, J., Massey, L.D., Mutter, S.A.(1988) Introduction. In: Intelligent tutoring systems: Lessons Learned. pp. 1-14 London: Lawrence Erlbaum Associates.

Van Lenthe, J., (1992) The development and evaluation of ELI, an interactive elicitation technique for subjective probability distributions. In: D. Leclercq, J. Bruno (eds.) Item Banking: interactive testing and self-assessment, NATO ASI series F, Vol. 112. Berlin: Springer-Verlag (this volume).

Zeiliger, R. (1991) SHIVA: manuel de référence, Delta D1010, 1-115. Ecully: CNRS-IRPEACS 93 Ch. des Mouilles PB 167, F-69131 Ecully Cedex, France.

Using Interactive Videodisc for the Assessment of Adult Learning Styles

John A. Gretes

Department of Curriculum and Instruction, College of Education, University of North Carolina at Charlotte, Charlotte, North Carolina 28223, U.S.A.

Abstract: This paper is an account of the design, development and evaluation of the interactive videodisc Learning Style Survey, an instrument based on sections of the paper-pencil version of the Hill Cognitive Style Interest Inventory. Psychometric properties of the instrument are discussed, including face, content, and construct validity as well as stability, equivalence, and internal consistency. Suggestions for the possible use of the instrument in the identification of learning style for the improved instructional success of low-literate adults are included.

Keywords: Adult literacy, Learning style, Low-literate adults, Interactive videodisc, Instrument validation.

1 Introduction

There have been several different views of learning and cognitive styles over the years. The German psychologist Carl Jung considered what he called "psychological types " as early as 1921. Others including Allport, Lowenfield and Klein (1951) identified "heptic types" and "visual types" , "levelers" and "shapers" and researched the impact of these types on learning.

The field focused on Personality models, Information Processing models, Social Interaction models, and Instructional-Preference models. Leaders in the Personality models included, Herman A. Witkin who pioneered work in the area of "Filed Dependence and Independence" and Isabell B. Myers, who developed the Myers-Briggs Type Indicator. Some other domains usually included with the Personality models are "Reflection versus Impulsivity" (Kagan 1965), The Omnibus Personality Inventory developed by the Center for Research and Development at the University of California at Berkeley in the 1950s, and the Holland Typology (Holland 1966) that has been used to determine environmental preferences. This model has examined personality traits that may influence learning.

The Information Processing models have developed based on the work of Pask (1975, 1976). Others contributing to the model include Siegel and Siegel (1965), McDade 1978), Ausubel (1963), Schmeck (1983), Kolb (1984). The Scoial-Interaction models have developed around the work of Mann (1770), Reichmann and Grasha (1974), Eson and Moore (1980), and Fuhrmann and Jacobs (Furhmann and Grasha 1983). The Information Processing models have looked at the processing sequences of individuals and how those sequences influence learning. The Social-Interaction models have examined the impact of social learning theory on the process of learning, and have attempted to define student to student and teacher to student interactions that can increase learning.

The Instructional-Preference Models, pioneered by Hill (1970) and Canfield (1980) have developed instruments to match student learning style with specific instructional methods to improve learning. Since these model also include focusing on learning how to learn they were the natural choice for use with research on low-literate adult populations in community colleges in the United States.

The identification of learning or cognitive style has been a documented practice with community colleges around the United States since the early 1970's. These community colleges use the cognitive or learning style information to help place adults into compatible instructional materials. The Hill (1970 & 1980) Cognitive Style Interest Inventory (CSII) is currently used by a consortium of some 23 member colleges in the League for Innovation in the Community College. The Hill instrument has been used and validated by Nunney (1978), Lange (1979), and Hand (1978) and according to Neil (1975) had been implemented in many programs from kindergarten through higher education. Many have reported on the use of the CSII and other instruments used to measure cognitive or learning style including Leclercq and Boskin (1990), Claxton and Murrell (1987), Archinlega & Arrigo (1974), Cargo (1982), Frizzell (1984), Hand (1978), Mickler & Zippert (1987), and Whitley (1982, May). Using the paper-pencil Cognitive Style Interest Inventory with low-literate adults presents some problems since the student must read and respond to a series of written items.

1.1 The learning style survey

The LSS program is based on the Cognitive Style Interest Inventory, a widely used learning style assessment among community colleges. Since this original instrument was a paper-based test, the project team was faced with the challenge of converting it to a valid videodisc-based format. In the early stages of development, an exhaustive audience analysis was completed to provide the design team with information on demographics, entry behaviors, and a wide variety of learner considerations. Members of the project team then analyzed the questions from the Cognitive Style Interest Inventory in terms of the reading level and produced a set of questions that could be easily understood by an individual reading at the fourth grade level. These thirty-two questions were the basis of the treatment plan. Exhibit 1 below displays a sample of the question match and examples of the questions for both the LSS and the CSII. When the initial treatment of the assessment instrument was finalized, a paper-based storyboard was produced. The storyboard began as a single "frame" for each of the thirty-two questions. The picture in each frame was designed to help the student visualize the question it helped represent. From these initial frames, the project team developed a series of short, sequential scenarios that lead to each question. At that time a complete storyboard was produced of all the scenarios using nearly one hundred black and white photographs.

The storyboard (a 3-ring binder containing over 40 pages of photographs and questions) was presented to fifty adult literacy students to try out the design. This tryout provided essential information about the face validity of the questions and the approach. The correlation between the paper-based Cognitive Style Interest

Inventory and the new storyboarded questions was calculated after a significant percentage of students were tested. Initial correlations were low (.34). The project team continued to monitor student responses to the questions and made revisions according to that feedback. After three rounds of revision, the new questions with the revised scenarios correlated to the Cognitive Style Interest Inventory at .84, showing that the videodisc script and the questions in the script were not significantly different from the original instrument. Once this satisfactory level of correlation was achieved, the video production process began.

The LSS Videodisc program begins with a short, highly motivational introduction to the concept of individual learning styles presented by Wally Amos (Songer & Gretes, 1990). He explains why understanding this information about oneself can lead to a more successful and rewarding learning experience. The user is then introduced to the main characters of the program and information about how to complete the survey. This introductory information is presented in a series of exercises designed to give the user an opportunity to practice interacting with the technology. After completing a practice question, the user begins the survey. Thirty-two questions are embedded in short (15-25 seconds) scenarios as the program follows six people through a typical day. The user has the option to repeat the question as often as necessary before answering.

After answering all the questions, the learner is presented with information that describes their preferred learning style. This feedback section is designed to provide specific information about how to learn new reading or math skills most efficiently.

Several example questions from the LSS and the CSII are presented in Exhibit 1 below. The sample provided in the exhibit does not include all 32 items. It does provide one or more examples for the AL-Auditory Linguistic, VL-Visual Linguistic, AQ-Auditory Quantitative, and VQ-Visual Quantitative aspects of style identified by Hill (1970 & 1980). Detailed accounts of the LSS to CSII item match are provided by Gretes and Songer (1988).

Exhibit 1
Sample Video and Paper-Pencil Item Match By STYLE

RESPONSE PATTERN CHOICES
R=Rarely S=Sometimes U=Usually
LEARNING STYLE
AL=Auditory Linguistic - VL=Visual Linguistic
AQ=Auditory Quantitative - VQ=Visual Quantitative

Video Items	Paper-Pencil Items
(AL)	
1. Do you try to listen to the radio?	9. I make it a point to listen to the news on the radio.
(VL)	
2. If you ask someone to write something down, do you read it to make sure it's right?	28. After I dictate a letter, I read it to be certain it is correct.
(AL)	

3. Do you understand things
better after you talk
about them?
(VL)

6. I do best on a test if it
covers information I have
discussed.

4. Are your written messages
easy to understand?
(AQ)

5. My written explanations are
easily understood.

5. Do you talk about price with
others before you buy
something?
(AQ)

15. I discuss "sale" prices with
others before I go shopping.

6. When someone talks to you
about numbers, is it easy
for you to understand
what they mean?
(VL)

22. I like verbal (oral) tests in
mathematics.

7. Do you use a map when you go
new places?
(VL)

11. I refer to or read a map when
I am going to a strange place.

8. Do you like to learn new
things by reading about them?
(VQ)

26. I prefer to acquire information by reading about it.

9. Do you write a telephone
number down to remember it?
(AL)

17. I write a telephone number
down to remember it.

10. When you go to a new place,
do you ask for directions?
(AL)

12. I prefer verbal directions for
finding a strange place.

11. Do you talk to your friends
on the phone?

10. I communicate with friends and
colleagues by telephone.

---

2 The validation study

2.1 Method

For the validation study, more than 500 community college students in seven states (North Carolina, Ohio, Oregon, Arizona, Florida, Illinois, California, and Missouri) were used. These subjects were entering community college students enrolled in Adult Basic Education (ABE) or English as a Second Language (ESL) programs. They provided a sample of the target population ranging in age from 16 to over 55, with reading ability from 4th grade to 12th grade level. The male subjects accounted for 39% of the sample, females were 61%. By race, there were 24% Whites, 43% Blacks, 27% Hispanics, and 6% Asians. Of the 500 students in the sample, usable data sets were obtained from 316. Subjects were asked to respond to the 32 item LSS and the CSII. On the LSS the subjects responded to each item scenario using a Sony interactive videodisc View System. On the CSII the subjects responded to the paper/pencil version. The scores based on the item responses were combined to yield scores for each of the four styles identified by Hill (1970 & 1980) as "theoretical symbols". These symbols are described by Hill as (AL) Auditory Linguistic, gaining meaning from words heard; (AQ) Auditory Quantitative, gaining meaning from numbers heard; (VL) Visual Linguistic, gaining meaning from words seen; and (VQ) Visual Quantitative, gaining meaning from numbers seen. The subjects were randomly assigned to one of three groups. In group one, subjects took the LSS and 3 to 5 days later the CSII. The subjects in group two took the CSII and 5 to 10 days later the CSII. In group three, the subjects took the LSS and 5 to 10 days later took the LSS again.

2.2 Results

Gretes and Songer (1990) reported an average correlation of equivalence based on subscale correlations between the LSS and the CSII of .60. The subscale correlations ranged between the low of .39 to the high of .69. Test-retest reliability for the CSII by subscale ranged from a low of .57 to high of .75. For the LSS, test-retest reliability by subscale ranged from a low of .69 to a high of .78. Internal consistency using Cronback's ALPHA for the LSS subscales ranged from a low of .82 to a high of .87. For the CSII by subscale, internal consistency using ALPHA ranged from a low of .78 to a high of .85.

Factor analysis was conducted using the <u>Statistical Package for the Social Sciences</u> subprogram Factor. Table 1 identifies loadings by factor for each of the two identified factors for the LSS and the CSII (Gretes & Songer, 1989 & 1990).

Table 1   LSS and CSII Factor Loadings and Correlations

<u>Factor Loadings</u>

N=195

	LSS Factor 1 (Visual)	LSS Factor 2 (Auditory)	CSII Factor 1 (Visual)	CSII Factor 2 (Auditory)
Auditory Linguistic (AL)	-.05	.56	-.07	.52
Auditory Quantitative (AQ)	.24	.71	.34	.63
Visual Linguistic (VL)	.52	-.02	.55	-.05
Visual Quantitative (VQ)	.84	.08	.82	.09

<u>Factor Correlations</u>	LSS Factor 1	LSS Factor 2		CSII Factor 1	CSII Factor 2
Factor 1	1.00	.49	Factor 1	1.00	.34
Factor 2	.49	1.00	Factor 2	.34	1.00

Table 2 displays a multitrait/multimethod matrix for the LSS and CSII by subscale (Gretes & Songer, 1989 & 1990). The factor pattern correlations and factor loadings for both the LSS and the CSII seem to support the idea that both instruments represent a single construct, namely cognitive style, with two factors identified as Auditory and Visual. It is recommended that any comparisons made using the LSS scores should be based on the four style or subscale scores, given the stability and internal consistency reported. The AL and AQ subscale scores for the LSS seem to combine as subsets of the AUDITORY factor while the VL and VQ subscale scores seem to combine as subsets of the VISUAL factor. For this reason, comparisons made using the LSS should be based on the two factor and/or the four subscale scores (Hand, 1978).

Table 2 Multitrait/Multimethod Matrix

N=195	METHOD 1 LSS				METHOD 2 CSII			
var	07 (AL)	08 (AQ)	09 (VL)	10 (VQ)	11 (AL)	12 (AQ)	13 (VL)	14 (VQ)
METHOD 1 -LSS								
var.07 (AL)		1.00	.43	.11	.22	.58	.36	.07	.23
var.08 (AQ)		.43	1.00	.27	.56	.28	.39	.20	.38
var.09 (VL)		.11	.27	1.00	.44	-.06	.21	.61	.45
var.10 (VQ)		.22	.56	.44	1.00	.09	.48	.30	.69
METHOD 2 - CSII								
var.11 (AL)	.58	.28	-.06	.09	1.00	.35	.04	.13
var.12 (AQ)	.36	.39	.21	.48	.35	1.00	.26	.53
var.13 (VL)	.07	.20	.61	.30	.04	.26	1.00	.45
var.14 (VQ)	.23	.38	.45	.69	.13	.53	.45	1.00

Table 3 National Field-Trial Percentage of Auditory and Visual Learners by Reading Level

Reading Level	% Auditory	% Visual	Total % of Sample
Elementary (4, 5, 6)	60	40	17
Junior High (7, 8, 9)	57	43	47
High School (10, 11, 12)	40	60	36
	51	49	100

Table 3 displays the results of a cross-tabulation in percentages between auditory and visual learners by reading level. Overall, 51% of the sample were identified as auditory learners. Of those reading at the elementary school level, the majority (60%) were identified as auditory learners. One possible explanation for the 60% reading at the elementary school level being identified as auditory learners is that these learners may have learned to rely on hearing rather than reading due to their problems with the written word. The majority of learners reading at the jr. high level were identified as being auditory learners (57%), while 43% were identified as visual learners. The majority of learners reading at the high school level were identified as visual learners (60%) with 0% identified as auditory learners. These learners reading at the high school level were much better readers than the two other groups and it is speculated that they have learned to rely on their improving ability to read in order to learn.

2.3 Conclusions

The following conclusions are drawn from the data reported above based on the sample representing low-literate adult learners served by Community College Adult Basic Education Programs in the USA:

1. The Learning Style Survey is stable over time base on the Pearson correlation of test-retest reliability of .78.
2. Based on the Cronback's ALPHA of .85, the LSS seems to be internally consistent.
3. The correlation of equivalence of .68 provides evidence that the LSS and CSII are equivalent forms.
4. The stability (.78) and the internal consistency (.87) of the LSS are slightly higher than those reported for the CSII, .75 and .85 respectively.
5. The LSS clearly measures two distinct factors identified as Auditory and Visual Learning Style.
6. Evidence of construct validity is provided based on the results of the factor analysis and the multitrait/multimethod matrix. That is, the LSS and the CSII seem to be measuring the same constructs.
7. Face and Content validity evidence are provided by the logical one to one match between LSS and CSII items.
8. Students in the sample who read at the elementary level seem to be more auditory than visual in learning style.
9. Students in the sample who read at the jr. high level seem to be more auditory than visual in learning style.
10. Students in the sample who read at the high school level seem to be more visual than auditory in learning style.
11. Whites in the sample are more visual than auditory and more linguistic than quantitative.
12. Blacks in the sample are more auditory than visual and more linguistic than quantitative.
13. Hispanics in the sample are split evenly in auditory and visual styles and more linguistic than quantitative.
14. Asians in the sample are more visual than auditory and more linguistic than quantitative.
15. Reading level seems to influence auditory/visual but not linguistic/quantitative dominance.
   - At all reading levels among all races in the sample, students were more linguistic than quantitative.
   - Auditory/Visual dominance varied among all races in the sample by reading level.
16. Students in the sample preferred taking the LSS over the CSII.

3 The future

If in fact learning styles differ, and there are interpersonal and intra-personal differences in learning styles (Leclercq & Pierret, 1989) then should the most efficient instrument for the determination of learning style allow the learner the choice of reporting their style using a multiple-media format? The answer to this question is most likely *yes*. In such a new instrument, the learner might select from several icons available on each computer screen. The icons might include Text, Audio, Graphics, Still Video, and/or Motion Video. The learner could select the mode of delivery or even multiple modes of delivery. Such computer software would verify the learner style by determination of paths taken to the items, the type of items and support selected, responses to the items, and validate the learner predictions of learning style.

More research needs to be conducted using software similar to that described above. Such research should not be limited to low-literate adults. It should be conducted with learners at all levels from preschool through graduate school.

We do see future research involving the the impact of learning style, as measured by the LSS, on student expressions of confidence. There does seem to be a logical connection between style and confidence since they can both change with instruction.

The integration of such learning style instrumentation with interactive training materials would seem to be the next logical step. Some specific standards for courseware development including learner selection options based on learning style might prove to be a very productive line of future research.

References

Allport, G. (1961). Pattern and growth in personality. New York: Holt, Rinehart & Winston

Archinlega, M., & Arrigo, A. (1974). A theoretical analysis of Hill's cognitive style inventory: Implications for assessing Mexican-American's preferred learning style. Journal of Instructional Psychology, 8 (1), 2-9

Ausubel, D.P. (1963). The psychology of meaningful verbal learning. New York: Grune & Stratton

Canfield, Albert. (1980). Learning styles inventory manual. Ann Arbor, Mich.: Humanics Media

Cargo, M. (1982). Research in learning style and reading. Theory into practice, 23 (1), 72-76

Claxton, C.C., & Murrell, P.H. (1987) Learning styles: Implications for improving educational practices. College Station, Texas: Association for the Study of Higher Education

Eison, J., and Moore, J. (1980). Learning styles and attitudes of traditional age and adult students. Paper presented at the 88th Annual Convention of the American Psychological Association, Montreal, Quebec, September

Fuhrmann, B., and Grasha, A. (1983). Designing classroom experiences based on student styles and teaching styles: a practice handbook for college teaching. Boston: Little, Brown & Co.

Frizzell, R. L. (1984). The status of learning styles. The Educational Forum, 303-312

Gretes, J.A., & Songer, T. (1990) The learning style survey (LSS), An interactive videodisc instrument: An instrument validation study, in Proceedings of the Seventh International Conference on Technology and Education, Brussels, vol. 2, pp. 435-437

Gretes, J.A., & Songer, T. (1989) Validation of the learning style survey: An interactive videodisc instrument, Educational and Psychological Measurement, Vol. 49, 235-241

Gretes, J.A., & Songer, T. (1988) Central Piedmont community college learning style survey (lss) interactive videodisc project instrument validation report, Charlotte, North Carolina: The READY Project

Hand, J.D. (1978). Educational cognitive style for health science students, final report: Part 2, Research design and analysis of data. Bethesda, Maryland: National Medical Audiovisual Center

Hill, J.E. (1970). Cognitive style as an education science. Bloomfield Hills, Michigan: Oakland Community College Press

Hill, J.E. (1980) The educational sciences: a conceptual framework. West Bloomfield, Michigan: Hill Educational Sciences Research Foundation

Hill, J.E. & Nunney, D.N. (1974). Personalizing educational programs utilizing cognitive style mapping. Bloomfield Hills, Michigan: Oakland Community College Press

Holland, J.L. (1966). The psychology of vocational choice. Waltham, Mass.: Ginn & Co.

Kagan, Jerome. (1965). Reflection impulsivity and reading ability in primary grade children. Child Development 36: 609-28

Kolb, D.A. (1984). Experimental learning: experience as the source of learning and development. New York: Prentice-Hall

Lange, C.M. (1979). Identification of learning styles. New York: National League for Nursing

Leclercq, D. (1990) A fresh look on confidence making, in Proceedings of the Seventh International Conference on Technology and Education, Brussels, vol. 1, pp. 646 - 649

Leclercq, D. & Boskin, A. (1990) Note taking behaviors studied with the help of hypermedia, in Proceedings of the Seventh International Conference on Technology and Education, Brussels, vol. 2, pp. 16-19

Leclercq, D. & Pierret, B. (1989) A computerized open learning environment to study interpersonal variations in learning styles: DELIN, in Proceedings of the Sixth International Conference on Technology and Education, Orlando, vol. 2, pp. 268-272

Mann, R.D., et. al. (1970). The college classroom: conflict, change, and learning. New York: John Wiley & Sons

McDade, C. (1978). Subsumption versus educational set: Implications for sequencing instructional materials. Journal of Educational Psychology, 70: 137-41

Mickler, M.L. & Zippert, C. P. (1987). Teaching strategies based on learning styles of adult students. Community/Junior College Quarterly of Research and Practice, 11(1), 33-37

Neil, M. (1975). Cognitive style: A new aspect of instructional technology, New Directions for Community Colleges, 3, 73-80

Nunney, D.N. (1978). Educational sciences dissertation abstracts. Bloomfield Hills, Michigan: American Education Sciences Association

Pask, G. (1976). Styles and strategies of learning. British Journal of Educational Psychology, 46: 128-48

Reichmann, S., and Grasha, A. (1974). A rational approach to developing and assessing the construct validity of a student learning styles scale instrument." Journal of Psychology, 87: 213-23

Siegel, L., and Siegel, L.C. (1965). Educational set: a determinant of acquisition."
Journal of Educational Psychology, 56: 1-12

Songer, T., & Songer, T. (1992) Using interactive videodisc for assessment of
adolescents and adults: the learning style survey, in Proceedings of the Seventh
International Conference on Technology and Education, Brussels, vol. 2, pp. 575-
577

Whitley, J. B. (1982). Cognitive style mapping: Rationale for merging the "old" and
"new" technologies. Educational Technology, 26-28

In Pursuit of the Fallacy: Resurrecting the Penalty

Emir Hamvasy Shuford

The Knowledge Group, P.O. Box 25102, Dallas, Texas 75225, USA

The most useful piece of learning for the uses life is to unlearn what is untrue.
Antisthenes (445-365 B.C.)

It's not what you don't know that hurts you. It's what you think is so that isn't.
Mark Twain

Beware of false knoweldge: it is more dangerous than ignorance.
George Bernard Shaw

As some of you know, I have been exploring the elicitation of probabilities off and on for almost 30 years. While there are many interesting and practical aspects of this application of the mathematics of probability, the one alluded to in these quotations escaped me for many years. I now understand, as does Prof. Darwin P. Hunt, that conventional methods of testing and training fail to detect the existence of a fallacy. To establish a fallacy one must assert something to be true that is in fact false. Objective testing does not involve such an assertion. In eliciting probabilities the resulting probability distribution immediately reveals whether the person considers an untruth to be true.

In the United Stades of America, objective testing and response methods have dominated much of education and military and industrial training. What are the consequences of this policy? Is it allowing any fallacies to arise and to continue undetected? The answers are just beginning to come in. For example, I recently elicited probabilities for the job knowledge of well-trained employees, many of whom had been working on the job for a number of years. On average these people proved to be misinformed on one out of every five problems. Similar results are showing up for fourth and fifth grade pupils. It is too early to say to what extent correcting these fallacies will improve learning and job performance.

While the existence of fallacies may be more dramatic in the case of education and training it also may have an impact upon testing for selection and classification. To explore this, I used my interactive testing software as shown in Appendix A and described in Appendix B to elicit probabilities on ten different tests, primarily of job knowledge. Each person's amount of knowledge is measured by the quadratic scoring function as described in Shuford, Albert & Massengill (1966). Take this as the criterion of ability or achievement that we would like to predict.

Now that I had each person's probability distribution for each of the items on the test I was able to infer the action they would have taken when presented with each of these scoring procedures.

1. PICK=one point for a correct answer, lose nothing for an incorrect answer.
2. ACT-2=one point for a correct answer, lose two points for an incorrect answer.
3. ACT-4=one point for a correct answer, lose four points for an incorrect answer.
4. ACT-9=one point for a correct answer, lose nine points for an incorrect answer.

A little reflection reveals that the optimal strategy for

1. PICK is to choose an answer--one that is as least as likely as any of the others and never to skip the item. For the chosen answer, p is not less than 1/k where there are k foils.
2. ACT-2 is to choose an answer if its odds of being correct are two or more to one and to skip the item otherwise. If an answer is chosen, p is greater than or equal to .667.
3. ACT-4 is to choose an answer if its odds of being correct are four or more to one and to skip the item otherwise. If an answer is chosen, p is greater than or equal to .800.
4. ACT-9 is to choose an answer if its odds of being correct are nine or more to one and to skip the item otherwise. If an answer is chosen, p is greater than or equal to .900.

Given the action taken by each person on every problem and given the correct foil it is straightforward to calculate the total test score achieved under each testing method. In this manner I proceeded to produce scatter plots to show how well total test score reflects the ability criterion (see Figures 1 - 4). The positive relation between test score and criterion is evident in all cases except for the PICK score on TEST 2. ACT-2 shows a marked improvement over the conventional PICK scoring. ACT-4 is almost as goods as ACT-2. Things begin to fall apart again with ACT-9. These visual impressions are confirmed by the correlations shown in Table 1.

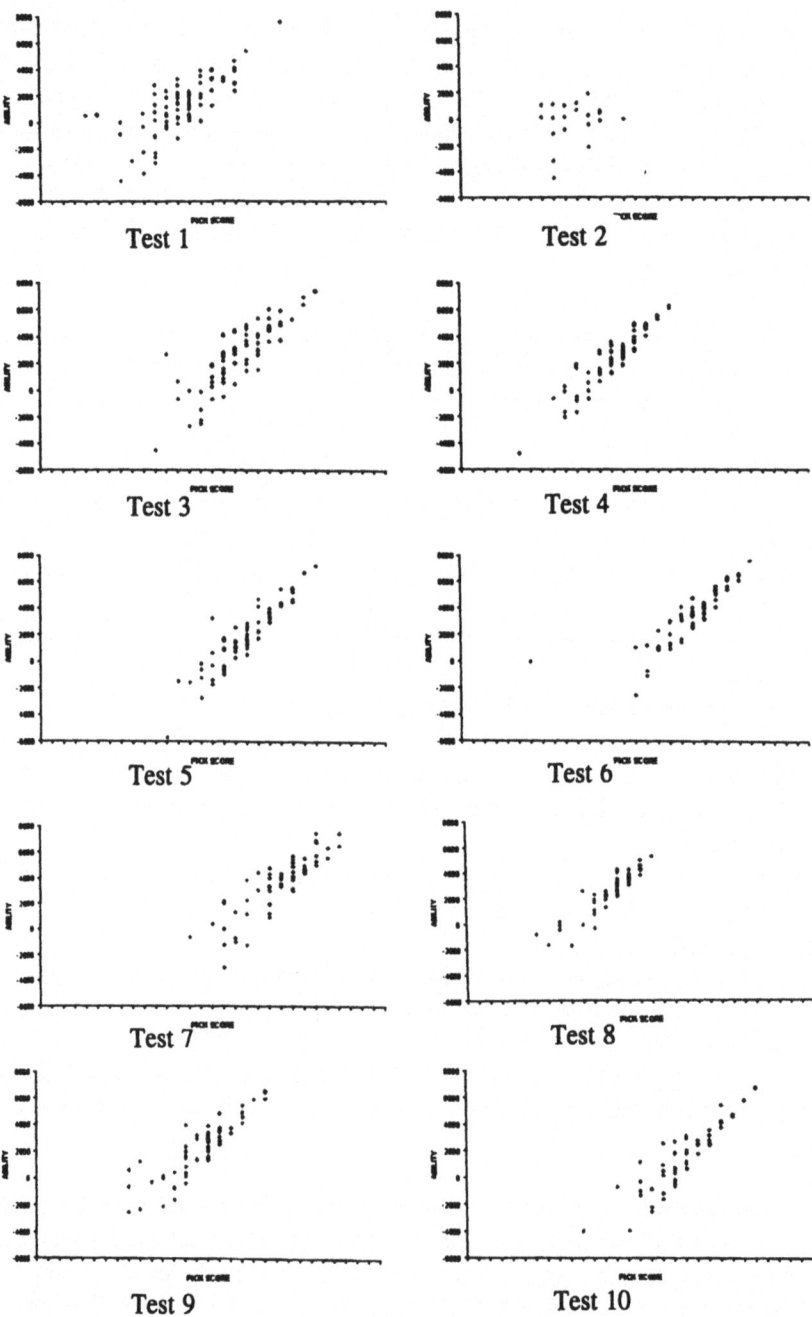

Figure 1. Ability or achievement as a function of PICK score for each of ten tests

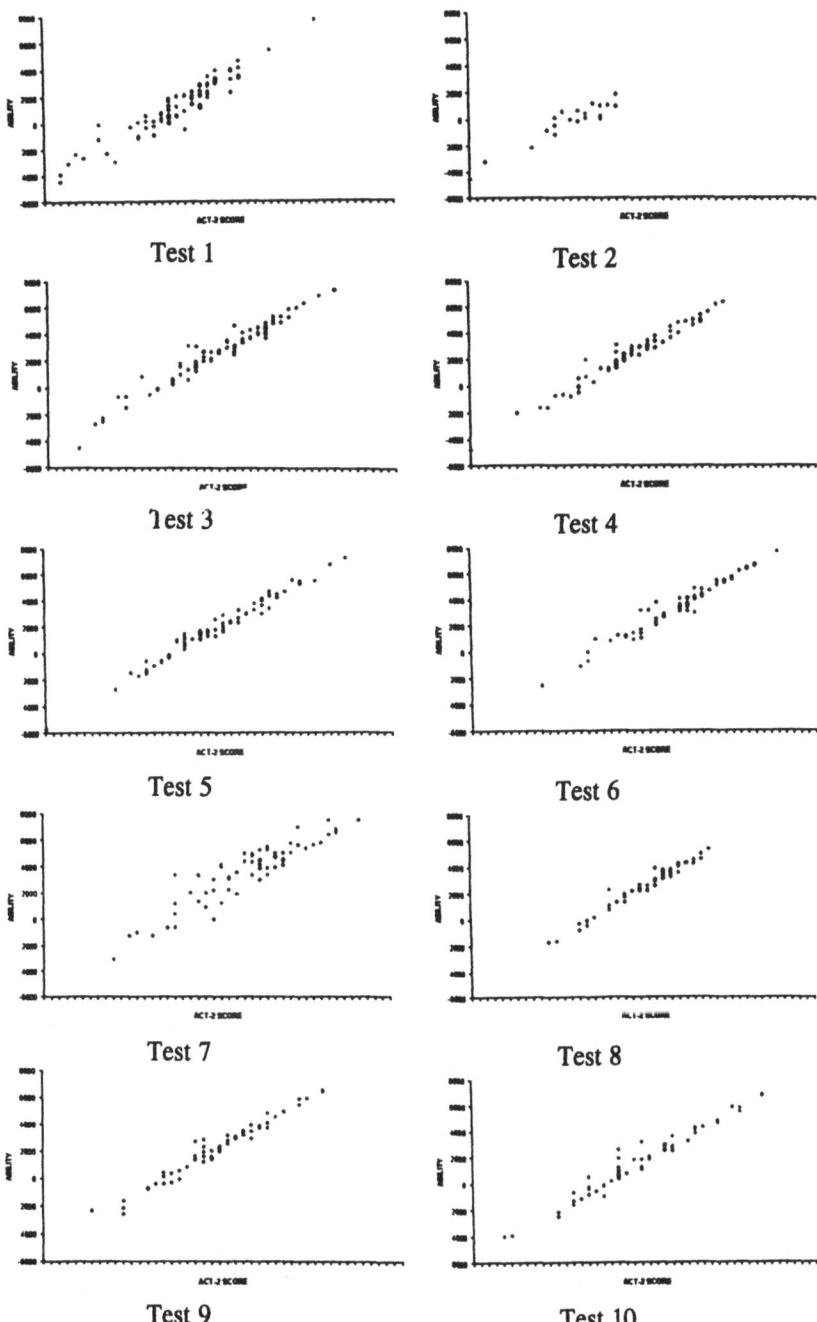

Figure 2. Ability or archievement as a function of ACT-2 score each of ten tests

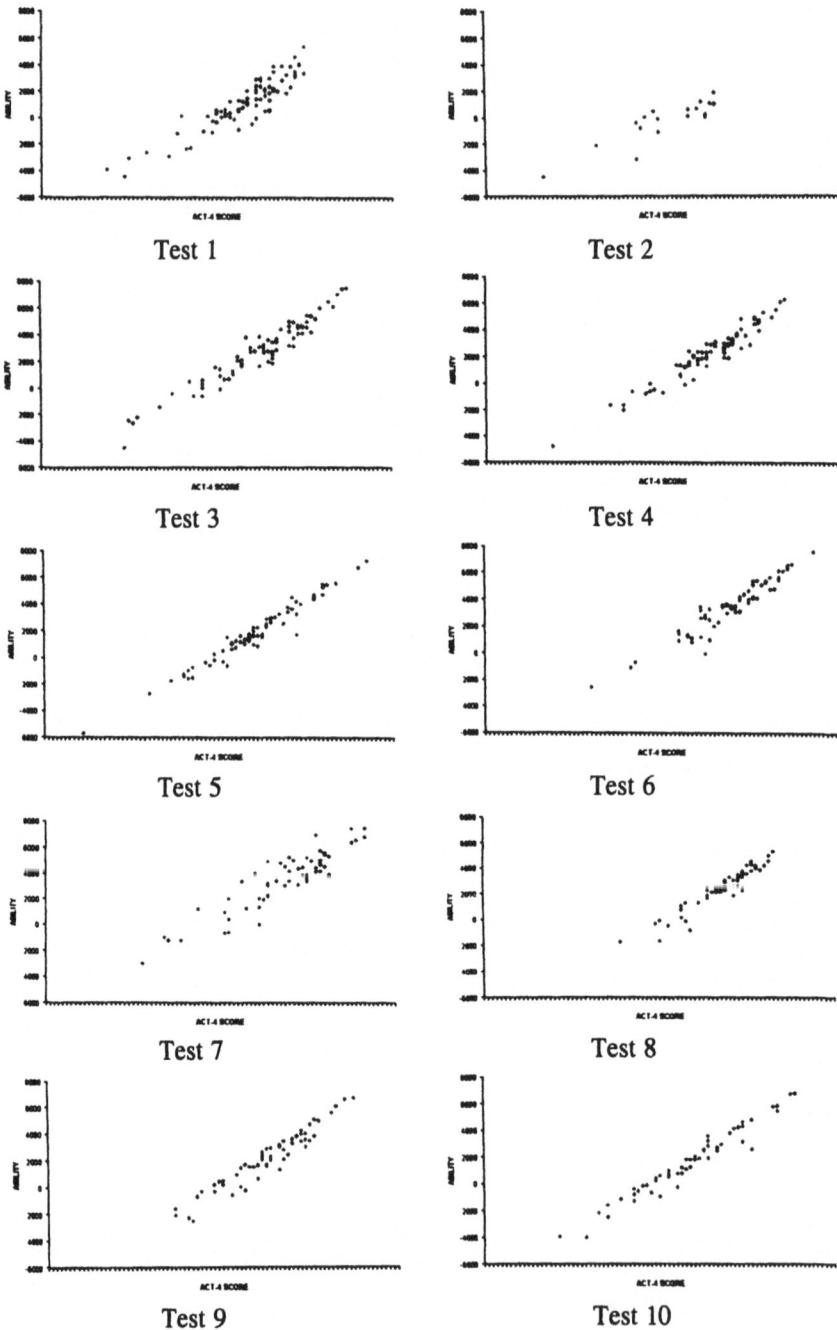

Test 1 Test 2

Test 3 Test 4

Test 5 Test 6

Test 7 Test 8

Test 9 Test 10

Figure 3 Ability or achieement as a function of ACT-4 score for each of ten tests

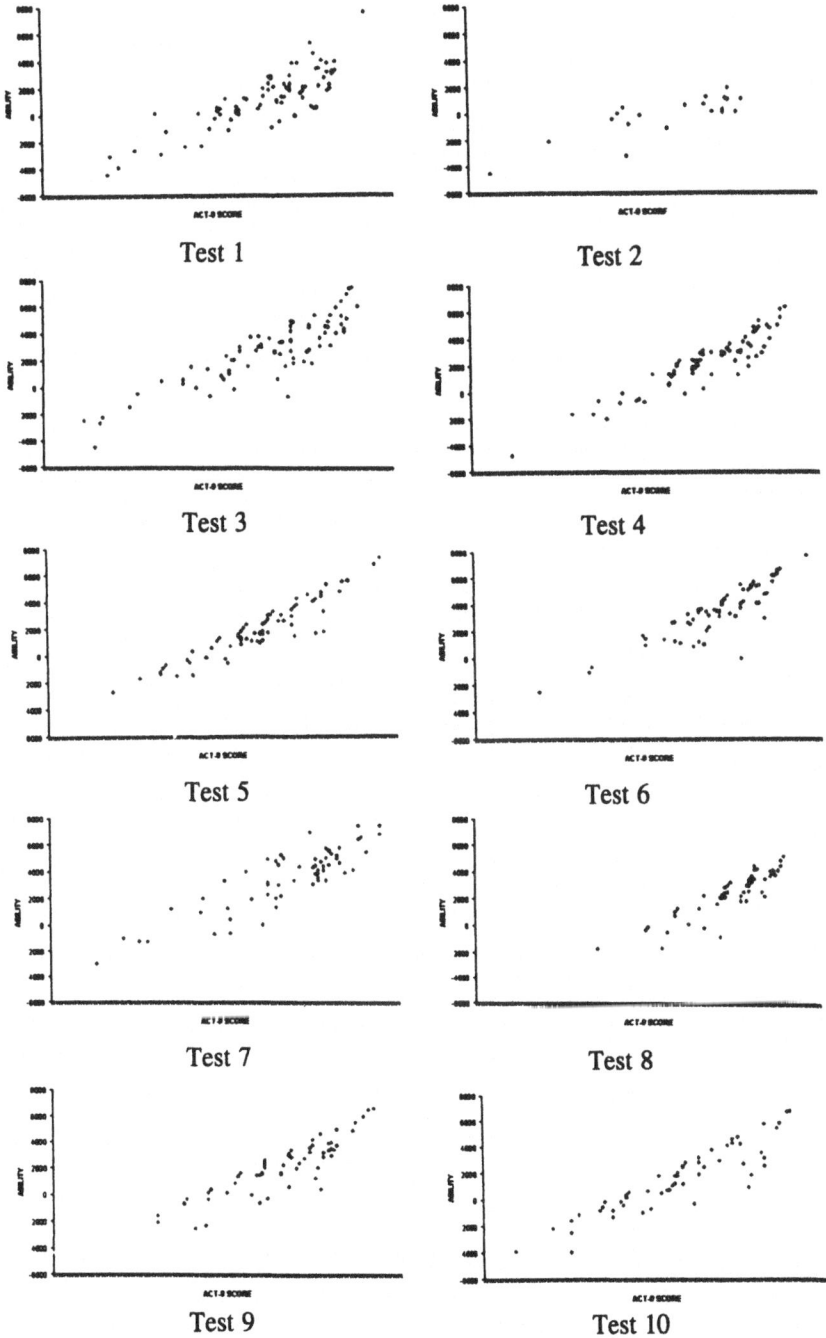

Figure 4. Ability or archievement as a function of ACT-9 score for each of ten tests

These correlation values may deceive by suggesting that the methods are very good and that the methods between them don't mean too much. The significance of these correlations depend upon the nature and importance of the selection or classification decisions to be made on the basis of these test scores. For example, consider what happens if you use these tests to award prizes to the very best of the group or to screen out the very lowest of the group as shown in Figure 5. While the PICK method is distinctly inferior to ACT-2, the best of the penalty methods, neither would be considered acceptable if they were being used to decide your fate.

An institution using these methods to make hiring, training, and promotion decisions might well be interested in the difference in mean ability yielded by a method as the cutting point is moved from minimum to maximum as shown in Figure 6. The differences between these 10 tests are considerable. ACT-2 generally performed very well and yielded separations for total test score close to the maximum possible. PICK performed less well, especially at the lower portions of the distribution. This probably reflects the noise introduced by the guessing strategies encouraged by PICK scoring. The practical importance of these performance differences between PICK and ACT-2 can only be determined, as always, from explicit consideration of the consequences of the decisions to be made using the test scores. Does the improved performance in each particular application offset the cost of changing from PICK to ACT-2 ? The answer depends upon the application but if it is really easy to change over why not do it everywhere ? Many of you may be thinking it can't be very hard to change from +1 and 0 scoring system to a +1, 0, and -2 scoring system. You may be right, but remember that if anything can be misunderstood and misused, people will usually manage to do it. So here are some cautions.

People need to understand what strategy to follow. They need to understand this prior to taking the test. It is not enough just to go back and rescore a test by subtracting two points for every wrong answer. This would just give a linear transformation of score and could not yield any additional information about anyone. ACT-2 gains precision because people do not select a foil in every instance. When they are facing considerable uncertainty they will not risk selecting an incorrect foil. They must understand this at the time they take the test.

People need to understand why that strategy should be followed. It won't do to tell people to follow a desired strategy and then later score their test in some manner that conflict with the advocated strategy. This may produce promising results for the first few experiments, but sooner or later people will find out that the strategy is not truly optimal and the method will break down.

Table 1. Correlation between score and criterion

TEST	# ITEMS	# PEOPLE	SCORING METHOD			
			PICK	ACT-2	ACT-4	ACT-9
1	22	80	0.68	0.94	0.91	0.83
2	22	19	0.29	0.94	0.88	0.81
3	25	77	0.85	0.98	0.96	0.87
4	18	75	0.92	0.98	0.96	0.92
5	24	76	0.91	0.99	0.98	0.94
6	25	71	0.87	0.98	0.96	0.89
7	27	61	0.87	0.92	0.92	0.87
8	16	59	0.92	0.98	0.95	0.89
9	20	61	0.86	0.98	0.96	0.89
10	26	58	0.91	0.98	0.98	0.91
	WEIGHTED	AVERAGE =	0.84	0.97	0.95	0.89

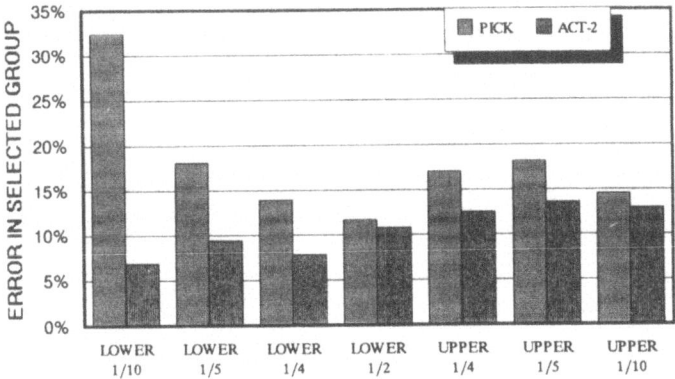

Figure 5. Percentage of people in selected group who do not belong there

Adding the penalty for selecting an incorrect foil causes people to think more while taking the test. For most ability, aptitude, and achievement tests this should be desired and help stabilize your results. There may be tests in use whose validity depend upon tapping very superficial and ephemeral things that are driven away by the thinking and reasoning encouraged by ACT-2.

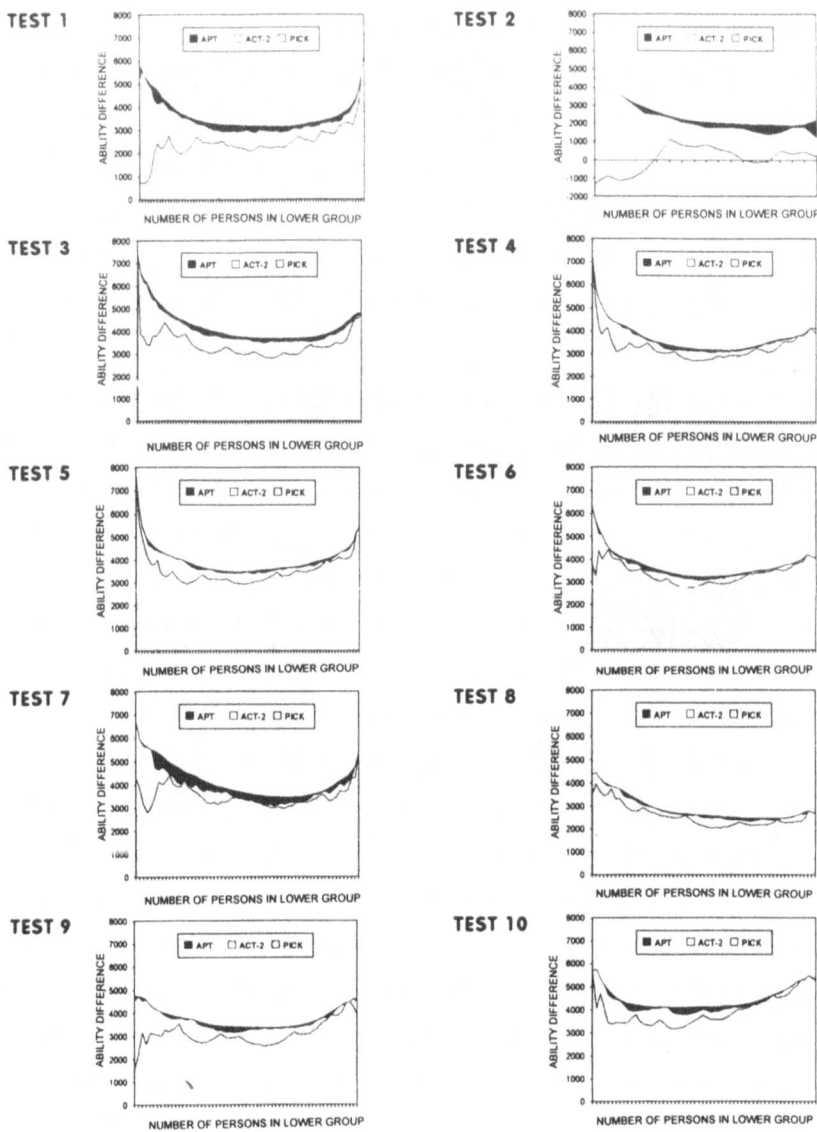

Figure 6. Difference in group means yielded by scoring systems.

REFERENCES

Shuford, E.H., Albert, A. & Massengill, H.E. Admissible Probability Measurement Procedures. *Psychometrika,* 31, 125 - 145 (1966)

APPENDIX A

Knowledge Program School for Jane Doe

The TOP line on this screen always shows which KEYS you should think about pressing.

Whenever you press a KEY not shown on this TOP line, a HELP BOX appears. The instructions you see may be different depending upon what you are doing.

Calling up and reading these HELP BOXES can guide you through this program. Also, you can learn strategies and tactics good for putting your knowledge to use.

He who knows not and knows not that he knows not,
  he is a fool, shun him.

He who knows not and knows that he knows not,
  he is a child, teach him.

He who knows and knows not that he knows,
  he is asleep, awaken him.

He who knows and knows that he knows,
  he is wise, follow him.

- Arabian Proverb

▶ ▶ ▶ ▶   W H I C H   W O U L D   Y O U   L I K E   T O   B E ?   ◀ ◀ ◀ ◀

```
F1=Help  F2=Log Off  F8=To History  Letter Key=Select Exercise  F9=Introduction
```

Key		Exercise Subject Matter..
A	0/3	Putting your knowledge to use
B	0/30	Understanding Computers 1
C	10/30	Understanding Computers 2
D	10/30	Understanding Computers 3
E	0/30	Understanding Computers 4
F	0/30	Microprocessors and Math Coprocessors
G	0/30	Memory, Displays, Printers, and Power
H	0/30	Using MS-DOS 1
I	0/30	Using MS-DOS 2
J	0/30	Using MS=DOS 3

```
F1=Help        F5=See Realistic        Letter Key=Select Exercise        F10=Exercises
```

Key	DOLLARS LOST BECAUSE OF YOUR LACK OF KNOWLEDGE OF THE SUBJECT									
	1st	2nd	3rd	4th	5th	6th	7th	8th	9th	10th
A	999	0	-	-	-	-	-	-	-	-
B	8,377	5,040	1,256	455	17	0	122	-	-	-
C	6,100	2,414	3,671	1,055	107	400	-	-	-	-
D	5,587	2,066	765	0	0	-	-	-	-	-
E	4,609	4,540	0	-	-	-	-	-	-	-
F	-	-	-	-	-	-	-	-	-	-
G	-	-	-	-	-	-	-	-	-	-
H	-	-	-	-	-	-	-	-	-	-
I	-	-	-	-	-	-	-	-	-	-
J	-	-	-	-	-	-	-	-	-	-

87

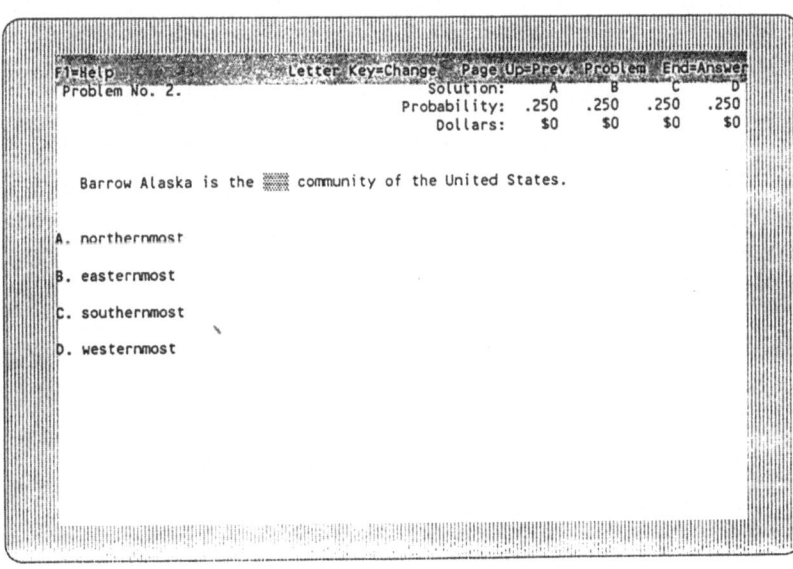

F1=Help	F6=See Knowledge	Letter Key=Select Exercise	F10=Exercises

DOLLARS LOST BECAUSE OF YOUR LACK OF REALISM

Key	1st	2nd	3rd	4th	5th	6th	7th	8th	9th	10th
A	0	0	-	-	-	-	-	-	-	-
B	4,288	1,577	105	10	0	0	3	-	-	-
C	3,501	640	815	65	2	0	-	-	-	-
D	16	72	12	0	0	-	-	-	-	-
E	751	0	0	-	-	-	-	-	-	-
F	-	-	-	-	-	-	-	-	-	-
G	-	-	-	-	-	-	-	-	-	-
H	-	-	-	-	-	-	-	-	-	-
I	-	-	-	-	-	-	-	-	-	-
J	-	-	-	-	-	-	-	-	-	-

F1=Help	Letter Key=Change	Page Up=Prev. Problem	End=Answer

Problem No. 2.

	Solution:	A	B	C	D
	Probability:	.250	.250	.250	.250
	Dollars:	$0	$0	$0	$0

Barrow Alaska is the ▨▨ community of the United States.

A. northernmost

B. easternmost

C. southernmost

D. westernmost

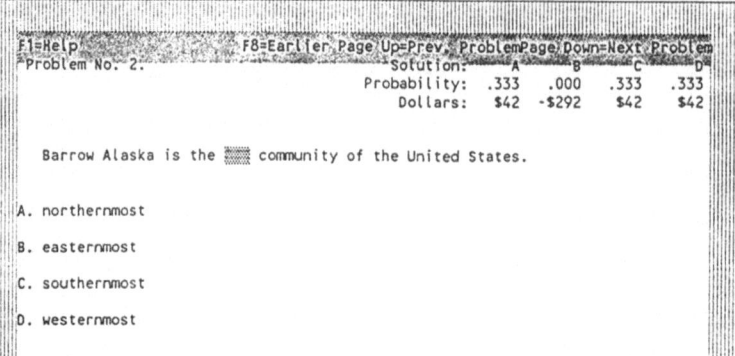

Probability: .333   .000   .333   .333
Dollars: $42 -$292 $42 $42

Barrow Alaska is the ▨▨ community of the United States.

A. northernmost

B. easternmost

C. southernmost

D. westernmost

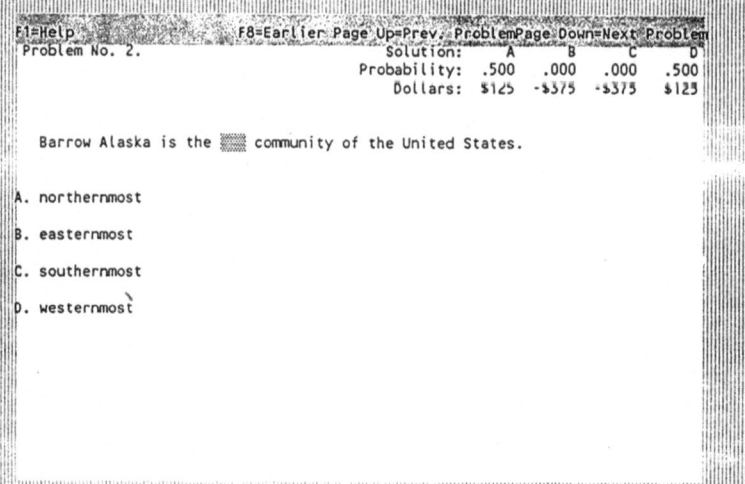

Probability: .500   .000   .000   .500
Dollars: $125 -$375 -$375 $125

Barrow Alaska is the ▨▨ community of the United States.

A. northernmost

B. easternmost

C. southernmost

D. westernmost

Problem No. 2.   Solution: A   B   C   D
                                        Probability: 1.000   .000   .000   .000
                                           Dollars: $373   -$625   -$625   -$625

    Barrow Alaska is the ▨ community of the United States.

A. northernmost

B. easternmost

C. southernmost

D. westernmost

Problem No. 2.   Solution: A   B   C   D
          INFORMED   Probability: .750   .000   .000   .250
    Solution A is the correct one.   Dollars: $312   -$438   -$438   -$188

    Barrow Alaska is the ▨ community of the United States.

A. northernmost

B. easternmost

C. southernmost

D. westernmost

EXERCISE: Computer Lab Quiz 2

TAKEN BY: Jane Doe   on August 31, 1992

STATE OF KNOWLEDGE - DIAGNOSIS	NUMBER	DOLLARS
Well-informed - ▬	8	$2,958
Informed - ▬	6	$1,642
Partially informed - ▮	2	$132
Uninformed -		
Misinformed - ▮	1	-$412
Badly misinformed -		
Total Dollars =		$4,320

Anticipated score: 86%
Actual score: 87%
You had a REALISTIC view of your knowledge of the Problems in the Exercise.

TA   ↔You can see here in the table all six possible States of Knowledge
─   and for each, the NUMBER of Problems on which you had that diagnosis,
      a SOLID BAR representing this NUMBER, and the number of dollars you
─   earned or lost with these Problems. These sums are added down at the
      bottom to give the Total profit or loss for the Exercise.
─   ↔By giving your Probability as to how likely you considered each of
      the possible Solutions to be the correct one, you allow the Program
─   to calculate earnings you expected to make at the time you answered
Pa   the Problems. It does this by multiplying each Probability by the
      corresponding dollars and summing these products over the Solutions
─   and then over the Problems.
─   ↔Both your Anticipated score and your Actual score are graphed near
      the bottom of the screen so you may see how realistic you are and how
─   you may improve to do better. Percentage numbers are also shown here.
B   ↔Select Earlier Status to see a Summary of previous knowledge.
      ↔Select Advice for an analysis of your performance with suggestions.
      ↔You can go back and study the Problems by selecting Review or
      you can select Exercises or Log (Off).
A ─────────────────→Press any key to erase this box.◄────────────────
      Actual score: 95%
You had a REALISTIC view of your knowledge of the Problems in the Exercise.

F1=Help F2=Log Off F6=Advice   F7=Current Status F10=Exercises   F5=Compare

EXERCISE: Computer Lab Quiz 2

TAKEN BY: Jane Doe   on August 30, 1992

STATE OF KNOWLEDGE - DIAGNOSIS	NUMBER	DOLLARS
Well-informed -	2	$750
Informed -	1	$280
Partially informed -	3	$67
Uninformed -	10	$2
Misinformed -	1	-$268
Badly misinformed -		
Total Dollars =		$855

Anticipated score: 68%
Actual score: 67%
You had a REALISTIC view of your knowledge of the Problems in the Exercise.

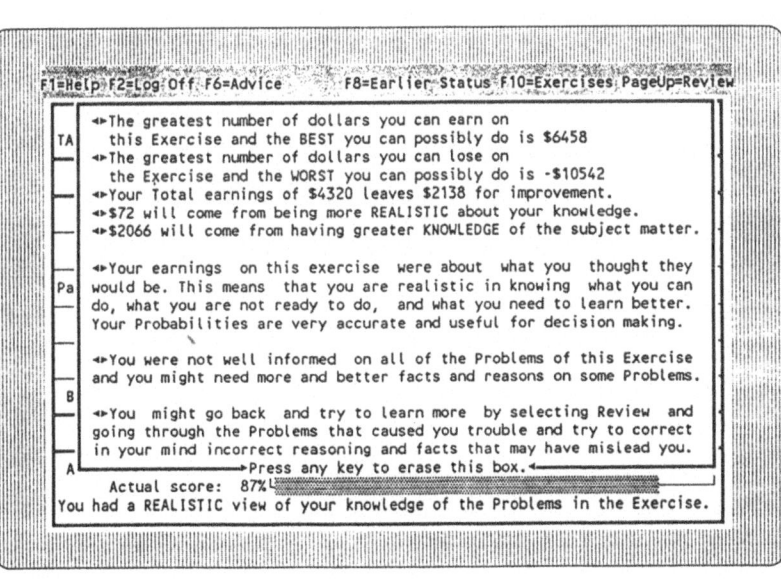

F1=Help F2=Log Off F6=Advice   F8=Earlier Status F10=Exercises PageUp=Review

TA
- The greatest number of dollars you can earn on
  this Exercise and the BEST you can possibly do is $6458
- The greatest number of dollars you can lose on
  the Exercise and the WORST you can possibly do is -$10542
- Your Total earnings of $4320 leaves $2138 for improvement.
- $72 will come from being more REALISTIC about your knowledge.
- $2066 will come from having greater KNOWLEDGE of the subject matter.

Pa
- Your earnings on this exercise were about what you thought they
  would be. This means that you are realistic in knowing what you can
  do, what you are not ready to do, and what you need to learn better.
  Your Probabilities are very accurate and useful for decision making.

- You were not well informed on all of the Problems of this Exercise
  and you might need more and better facts and reasons on some Problems.

B
- You might go back and try to learn more by selecting Review and
  going through the Problems that caused you trouble and try to correct
  in your mind incorrect reasoning and facts that may have mislead you.

A ──────────── Press any key to erase this box. ────────────

Actual score: 87%
You had a REALISTIC view of your knowledge of the Problems in the Exercise.

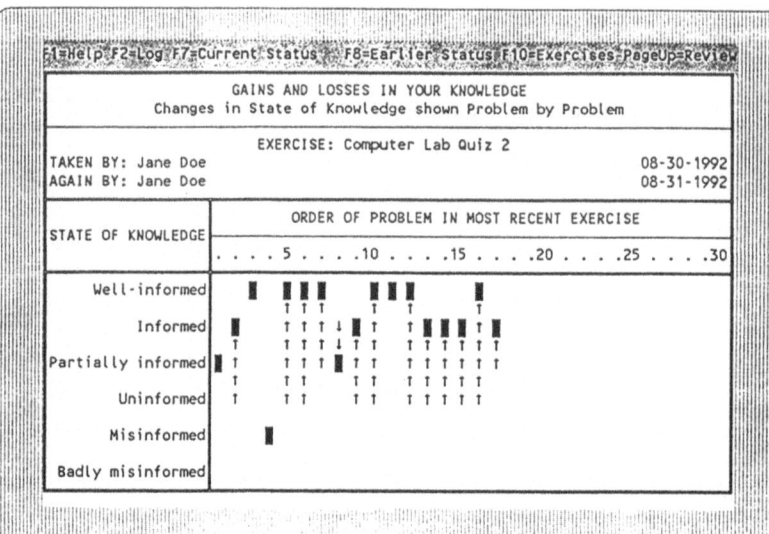

```
F1=Help F2=Log F7=Current Status   F8=Earlier Status F10=Exercises PageUp=Review
┌──────────────────────────────────────────────────────────────────────────┐
│                   GAINS AND LOSSES IN YOUR KNOWLEDGE                        │
│            Changes in State of Knowledge shown Problem by Problem           │
│                                                                            │
│                      EXERCISE: Computer Lab Quiz 2                          │
│   TAKEN BY: Jane Doe                                         08-30-1992     │
│   AGAIN BY: Jane Doe                                         08-31-1992     │
│                          ┌──────────────────────────────────────────────── │
│   STATE OF KNOWLEDGE     │     ORDER OF PROBLEM IN MOST RECENT EXERCISE     │
│                          │ . . . .5 . . . .10 . . . .15 . . . .20 . . . .25 . . . .30 │
│                          ├──────────────────────────────────────────────── │
│      Well-informed       │    ▌   ▌▌▌     ▌▌▌     ▌                         │
│                          │     ↑ ↑ ↑     ↑   ↑       ↑                      │
│         Informed         │ ▌   ↑ ↑ ↑ ↓▌↑ ↑   ↑ ▌▌▌▌↑▌                      │
│                          │ ↑   ↑ ↑ ↑ ↓↑ ↑   ↑ ↑ ↑ ↑ ↑                      │
│   Partially informed     │▌↑   ↑ ↑ ↑▐↑ ↑   ↑ ↑ ↑ ↑ ↑                       │
│                          │ ↑   ↑ ↑   ↑ ↑   ↑ ↑ ↑ ↑ ↑                        │
│        Uninformed        │ ↑   ↑ ↑   ↑ ↑   ↑ ↑ ↑ ↑ ↑                        │
│                          │                                                  │
│        Misinformed       │    ▐                                            │
│                          │                                                  │
│     Badly misinformed    │                                                  │
└──────────────────────────────────────────────────────────────────────────┘
```

```
F1=Help F2=Log F7=Current Status   F8=Earlier Status F10=Exercises PageUp=Review
┌──────────────────────────────────────────────────────────────────────────┐
│                   GAINS AND LOSSES IN YOUR KNOWLEDGE                        │
│     ┌────────────────────────────────────────────────────────────────┐     │
│  ── │ ◄►You can see here in this table  how your knowledge has changed from │ ── │
│     │ the previous time you did this Exercise.  You see the Problems in the │    │
│ TAK │ same order you saw them the last time you did this Exercise.  You may │ 2  │
│ AGA │ have seen them before in a different order.  Whoever is creating  the │ 2  │
│  ── │ Exercise can choose to scramble the order of Problems and Solutions.  │ ── │
│     │ ◄►You may  have taken  this Exercise  previously  before beginning to │    │
│ STA │ think about and to study  the subject matter of this Exercise.  Then, │ ── │
│     │ after finishing a module, chapter, or course of instruction,  you may │ 0  │
│  ── │ have done  this Exercise over again.  The changes  in this table show │ ── │
│ Ver │ how much you learned about each of the Problems from this experience. │    │
│     │ Lack of gain may be due to you, the Problem, or the instruction.     │    │
│     │ ◄►You may have taken the Exercise some time ago and now more recently │    │
│     │ in order to see how your knowledge of the subject  is holding up over │    │
│ Par │ time. Loss of knowlege is probably due to lack of use and forgetting. │    │
│     │ ◄►Select Current Status to see the Summary of current knowledge.     │    │
│     │ ◄►Select Earlier Status to see the Summary of previous knowledge.    │    │
│     │ ◄►You can go back and study the Problems by selecting Review or       │    │
│     │ you can  select Exercises or Log Off.                                 │    │
│     │ ───────────────►Press any key to erase this box.◄───────────────     │    │
│  Badly misinformed│                                                        │
└──────────────────────────────────────────────────────────────────────────┘
```

APPENDIX B

APTware 9.50

Software for the IBM PC, XT, AT, PS/2 and true compatibles.

Editor Program

- MAKE program and files needed to create and edit education programs.
  - Client logo and identification on initial screen.
  - Password needed for entry to program.
  - Bottom line on screen indicates which keys are active and what they do.
  - Context sensitive help screens activated by pressing F1 key or by hitting any inactive key.
- Introduction to each education program may contain any number˙ of screens.
- Menu for an education program produced by this program may contain up to 10˙ exercises.
- Exercises made up of any mixture of true-false and multiple-choice problems with up to five options may be created with this program.
- Each problem has one screen listing the options which may be preceded by up to 10˙ screens describing a scenario or otherwise presenting the problem.
- Each exercise may contain from one to 30˙ problems.
- The problem section of each exercise may be preceded by a prescript composed of up to 20˙ screens of introductory or background material.
- The problem section of each exercise may be followed by a postscript composed of up to 20˙ screens of explanatory or follow-up material.
- Exercises, introduction and menu are encrypted and ready for transfer to Student program.
- Allows choice of whether problems in the exercise are presented to students in the order shown or whether the

order of problems is randomly determined each time student begins the exercise.

- the options are presented to students in the order shown or whether the order of the options to each problem is randomly determined each time student does the exercise.
- students see the correct answer and the diagnosis of state of knowledge as the student does the exercise.
- whether student can go back to review prescript section.

- You can
  - create a new exercise or edit an existing exercise.
  - rename an exercise or change whether or not the order of the problems and options are scrambled and whether knowledge of results is given to the Student
  - insert, delete, change or add exercise problems. If all problems are deleted, the exercise itself is deleted.
  - insert, delete, change or add introductory, prescript and postscript screens.

- Exercise problems and other screens are made up of IBM PC symbol set graphics and text.

- You can print any of these screens and problems along with identifying information and date for documentation purposes.

Student Program

- TAKE program and files needed for up to 30** students within a subdirectory or on an individual student diskette.
  - Client logo and identification on initial screen.
  - Top line on screen indicates which keys are active and what they do.
  - Context sensitive help screens activated by pressing F1 key or by hitting any inactive key.

- For sign-on version, Last name (14 letters), and ID number (6 letters) elicited first time used and then an exact match required for all subsequent uses.

- If name and number is not an exact match (ignoring case) then student has a second chance to enter correctly.

- Teacher can enroll a new student at this time.

- Supports student doing objective exercises on-line using the new Applied Probability Technologgy

- Reads and decodes encrypted exercises produced using the parent Editor's program.

- Presents selected exercises with or without knowledge of results and with either a fixed or scrambled order of problems and options.

- When knowledge of results allowed, student can page back to examine any previously completed problem, but cannot change answer.

- If student has done exercise before and knowledge of results allowed, student can also examine how problem was answered previous time.

- Shows initial probabilities all equal for each problem and with no points to be earned nor lost.

- Allows student to modify probability for any solution by selecting solution and then dynamically increasing or decreasing new probability.

- Adjusts all the other probabilities so that the sum of all probabilities remains one to reflect fact that one of the solutions is correct or best.

- Adjusts the possible points earned or lost over a thousand point range according to the predictions made by the student.

- Allows any number of adjustments by student and provides a reset to equal (with zero gains and losses) key for use if needed.

- Reveals correct answer and interprets student's state of knowledge as well informed, informed, partially informed, uninformed, misinformed and FALLACY.

- Immediately records student's probabilities and response times for each question to protect against loss of data in the event of power failure.

- Allows student to leave exercise after completing any problem and reenter exercise later beginning with next problem.

- When student completes exercise, the TAKE program shows a summary screen which
  - may be printed by pressing the print screen key.
  - shows the student's states of knowledge and the contribution to total exercise score.
  - graphically displays exercise results to student in terms of total score anticipated by the student compared with the actual total score.

- when requested provides advice interpreting student performance on exercise.

- If student has done exercise before, student can
  - see this summary information for the time before.
  - see a problem-by-problem graphic display of how state of knowledge changed before and after. May be printed by pressing print screen key.
  - replay exercise comparing predictions before and after.
- Records student's responses on program for later analysis and review by teacher.
- Exercise Menu
  - contains up to 10** exercises for use by student.
  - shows for each exercise how many problems have been completed (when student has left exercise before completing all problems).
  - keeps track of how many times each exercise has been taken (up to 10 times** each).
  - allows student to do an exercise again at any time
- At Exercise Menu student may choose to see either of two History Tables
  - Loss in potential earnings due to lack of knowledge.
  - Loss in potential earnings due to lack of realism.
  - Each History Table can contain 100 entries—10 Repetitions by the 10 Exercises.

Class Program

- ENROLL program for initial enrollment of up to 30* students for an education program.
  - Password needed for entry to program.
  - Bottom line on screen indicates which keys are active and what they do.
  - Context sensitive help screens activated by pressing F1 key or by hitting any inactive key.
- If selected to be used, this program appears the first time the Student Program is run and then, after use, is deleted from diskette.
- Instructor can designate authorization code to be used with diskette and then enroll initial group of students.
- Instructor enters Last name (14 letters), and ID number (6 letters) for each of the student.

- Enrolled students displayed on screen during enrolling.

Teacher Program

- VIEW program and files needed to review student performance and progress.
  - Password needed for entry to program.
- Shows list of students and waits for a student to be selected and then shows exercise menu for that student and waits for an exercise to be selected.
- When selected and confirmed,
  - displays a summary table for the exercise showing student's
    - score,
    - states of knowledge, and
    - date and time of completion.
  - may be printed by pressing Print Screen key.
- When student has done the exercise before this time, displays also a summary table for the previous time showing student's
  - score,
  - states of knowledge, and
  - date and time of completion.
- When student has done the exercise before this time, can produce graph showing changes in state of knowledge problem-by-problem.
- Summary table may be printed by pressing print screen key.
- Review of this comparative graph shows any learning or forgetting by the student.
- Before and after comparisons can be printed by pressing print screen key.
- This replay mode allows teacher to proceed forward or backward through the exercise at will in order to support debriefing and review with the Student.

Problem Writer's Program:

- ITEM program containing data analysis procedures to assist in evaluating and revising exercise problems.
  - Client logo and identification on initial screen.

- Password needed for entry to program. Bottom line on screen indicates which keys are active and what they do.
  - Context sensitive help screens activated by pressing F1 key or by hitting any inactive key.
- Problem writer selects exercise for analysis from menu and whether data is to be accumulated or an analysis done of data collected to date.
- Can print a detailed report of student performance on exercise when data is accumulated.
- Collects data from all students on the program taking the same exercise on the specified occasion.
- Can accumulate results across a Student Program.
- Reads and decodes exercise results recorded on a Student Program produced by same parent Editor's Program and accumulates these data in one or more files.
- Assembles all the data in the correct order regardless of the sequence of presentation.
- Prints the question, options, and the key and then the prediction distributions for students ranked according to exercise score and according to problem score. Dissimilarities between graphs may indicate defect in problem.
- Prints an item summary showing distribution of states of knowledge for each item.
- Prints a class summary showing states of knowledge for students ranked according to total number of points earned.
- Programs can be provided to convert accumulated data files to comma separated variable format for input into EXCEL, and other spreadsheet and database programs.

* Note: Random access memory in target computer may limit how much text and other information can be placed in education program.

** Note: Size of the diskette memory may limit how many students can be enrolled and how many times they can do the exercises.

Subjective Uncertainty and the Structure of the Set of all Possible Events[1]

Jean-Marc Fabre

Université de Provence and CNRS, Centre de Recherche en Psychologie Cognitive, UFR de Psychologie et Sciences de L'Education, 29 avenue Robert-Schuman, 13621 Aix-en-Provence Cedex 1, France

Abstract: Two approaches to the concept of uncertainty are presented here as complementary methods. The first approach studies the assessment of certainty or accuracy probability in knowledge questionnaires. The issue in this case is the structure of the questions and the knowledge, and the difficulty subjects experience when building an adequate representation of the different perspectives used to understand the proposed choices. The results of several past and current studies are reported to illustrate the secondary nature of confidence ratings and the indicative role they can play. The second approach involves the analysis of how common expressions of uncertainty are coded. A recent finding is used to demonstrate that the representation of uncertainty depends on the makeup of the reference universe. When the choices which constitute that universe are clearly distinguished and delineated, certainty and uncertainty are two distinct dimensions. The difference between certainty and uncertainty is not so obvious, however, when the context in which a choice becomes meaningful is not clearly defined. In conclusion, it is proposed that the concept of realism in confidence assessment be reconsidered and newly viewed as the relationship between a subjective rating scale showing uncertainty at one end and certainty at the other, and a representation of the problem context underlying each question.

Keywords: Subjective uncertainty, confidence assessment, confidence rating scale, knowledge questionnaire

Introduction

The objective of this paper is to discuss the issue of the discrepancy between certainty and realism. Experimental data will be used to support two different points of view, one pertaining to knowledge questionnaires and confidence ratings, and the other, to the coding of the verbal expressions used to label the response scales.

---

[1] Translation due to Vivian Waltz.

The first point of view is based on the relationship between the structure of the questions asked and the structure of the knowledge entailed. In this case, the subject's confidence rating is treated as a secondary response, which is related in some way to the main responses given on the questionnaire (usually multiple choice or true-false questions).

The second point of view deals with the way in which uncertainty is expressed verbally, and with the relationship between the numbers on the response scales and the expressions used to label them.

Many studies have been conducted and published with one or the other of these two approaches. My intention here is to further this line of research, and show that it may be worthwhile to consider these two approaches as complementary.

The object of these studies has been to attempt to specify and clarify the concept of *uncertainty*.

We shall start with the following groundwork:

- When answering knowledge questionnaires, subjects make mistakes. The very purpose of asking them to give confidence ratings is to allow them to modulate each response by stating if and to what extent there is some chance that their response is incorrect. To assess their degree of certainty, subjects can proceed by processing all the available information on the question, recalling the outcomes of other comparable situations, or making a profit-loss calculation, etc. Externally, we can use a method which allows us to determine how realistic, not to say exact, the confidence ratings are, and we can then study the correlation between response accuracy and certainty assessment. Frequently, we find that there is a positive correlation between certainty and accuracy, and also that, in the absolute, the realism criterion is not met. In some cases we find overconfidence or underconfidence, which can even vary across questions within the same questionnaire. It is known that offering incentives such as rewards to trigger more realistic assessments is not sufficient. As the most common case found in the literature is overconfidence (Lichtenstein, Fischhoff, & Phillips, 1982), particularly when subjects are being questioned on general knowledge (semantic tasks; McClelland, Coulson, & Icke, 1990), the central problem posed for the psychologist is the assessment of uncertainty.

- In everyday life, as well as in teaching and experimentation, subjects spontaneously express states of uncertainty. They do so freely, using verbal or numerical expressions. Research and teaching applications rely on standard scales, which are usually simple enough so that even children can understand and use them without difficulty. However, as in all judgment activities, the assessment of uncertainty is not considered as the simple reflection of the feeling of uncertainty resulting from the processing of the information in the question. It is an activity in itself, with its own organization and its own operating laws. To demonstrate the specificity of this activity, experimental research is needed to determine what factors actually contribute to its variation.

The distinction can be made between at least two levels in uncertainty assessment: (1) the level at which the state of uncertainty is generated, and (2) the level at which it is expressed as a confidence rating. Different factors no doubt take effect at each of these levels and contribute to the feeling of uncertainty, whether in everyday situations or on standardized questionnaires. However, the two points of view I present here are complementary approaches to the *relationship* between the various meanings of uncertainty, and the cognitive context - the set of all possible events. A cognitive approach, in particular an approach that take into account recent results on context effects and cognitive biases, is necessary for a good appraisal of self-estimation techniques by teachers.

1 Knowledge Questionnaires

1.1 Link between decisions and meta-cognitive responses: the example of true-false questions

One initial point must be made here: the degree of confidence subjects assign to a given response is tightly linked to the popularity of that response. As a meta-judgment, a confidence rating is but one indicator of this popularity. Another indicator is the distribution of the main responses to the questionnaire, i.e. the answers chosen by the subjects. This can be assessed by studying the set of all responses made by a relatively homogeneous group of subjects. In my previous work (Fabre, 1980), I demonstrated that there was a correlation between the mean confidence rating which accompanies a given answer, and the frequency of occurrence of that answer.

It follows from this observation that the more a given population shares a point of view, the more confidence the population has in it, regardless of whether it is correct or not. Consequently, if an incorrect statement seems very believable, i.e. if it is so difficult that many errors are made, then these frequent errors will, on the average, be accompanied by a high degree of confidence. Hence, the commonly observed correlation between overconfidence and question difficulty, which in this case is a correlation between certainty and distractor attractiveness.

This result was obtained using true-false questions. For this type of question, we usually find a majority of correct responses, and a higher frequency of responses where "True" was chosen than the actual number of true statements. It follows (a) that correct responses are accompanied, on the average, by higher confidence ratings than incorrect responses; this is consistent with a common prejudice, but in reality can be explained by the same mechanism that accounts for overconfidence; and (b) that "True" answers are accompanied, on the average, by higher confidence ratings than "False" answers.

Table 1 and figure 1 (taken from Fabre, 1980) illustrate this observation.

Table 1
Percentage of responses "True" and "False" when statements are actually true or false, respectively (taken from Fabre, 1980, experiment B). Each cell in the table shows whether the response is correct (+) or incorrect (-), and the percentage of subjects who chose the corresponding answer.

answer statement	"True"	"False"
true	+ 87%	- 13%
false	- 36%	+ 64%

Figure 1

Affirmative true or false statements. Diagram of the correlate between type of response, ("True" or "False"), accuracy of response (correct or incorrect), and confidence ratings (on a three points scale, where 1 means low confidence and 3 means high confidence) (taken from Fabre, 1980 experiment B).

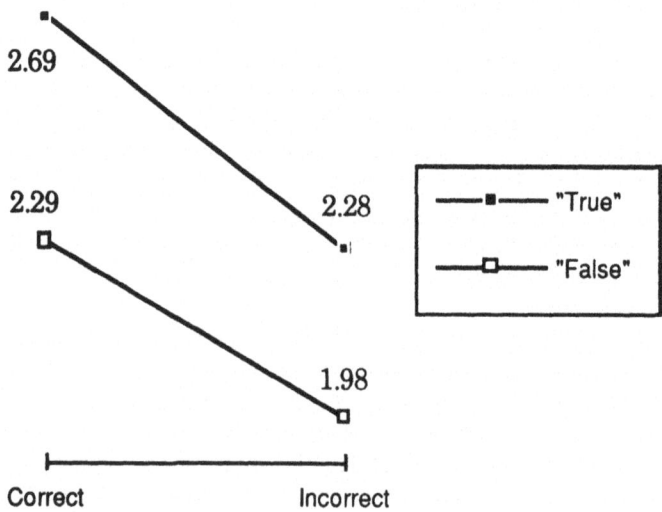

1.2 **Insufficient knowledge and judgmental processes: non realistic confidence assessment does provide specific information**

Even with true-false questions, the structure of the questions can be manipulated. This does not mean their formal structure, i.e., the rules for answer selection (Noizet & Fabre, 1975), but the structure of the knowledge _per se._

Take the case of a question on a true-false questionnaire consisting of a statement which expresses a relation between two items. In this case, the statement is true if and only if both items are true _and_ the relation is true, as in the example below.

Example 1
Two statements (out of five) taken from the coding hypothesis question
The sentence memorization coding hypothesis predicts specifically that:
(c) Active sentences will be memorized more easily than passive sentences .
(d) Sentences containing common words will be memorized more easily than sentences containing uncommon words.

The five statements share an item taken from the course ("the sentence memorization coding hypothesis") and a relation ("predicts specifically"). The second item is specific to each of them. Statements (c) and (d) were highly believable: 81% and 76% of the students judgent them trues, respectively. For this question as a whole, true statements (like c) and false but believable statements (like d) were accompanied by a higher degree of confidence when the answer was "True" than when it was "False", and the reverse was observed with unbelievable statements. See table 2.

Table 2
Coding hypothesis question. Mean ratings (on a three points scale, where 1 means low confidence and 3 means high confidence) of true statements, and believable or unbelievable false statements, when the subjects responses were either "True" or "False" (taken from Fabre, 1980).

	True statements	False statements	
Subjects reponses		believable	unbelievable
"True"	2.79	2.44	1.50
"False"	2.07	2.35	2.19

Subjects can determine whether the question is true or false in several different ways. They can simply ask themselves whether the items are true, without considering the relation between them. Let us assume that the subjects' knowledge of the question allows them to determine this correctly. In this case, if at least one of the items is false, the subjects will correctly respond "False". Or if both items are true and the relation is also true, they will correctly respond "True". However, if both items are true but the relation is false, they will *incorrectly* respond "True". Thus, the strategy which consists of basing one's response on the simple recognition of statements taught in the course, without considering the relation between them, is highly likely to trigger incorrect responses with a strong feeling of certainty.

In order to eliminate this overconfidence, the subject must use a different strategy. He or she must treat the problem posed by asking whether the relation between the two items is correct. In the coding hypothesis question (example 1), the problem to be solved requires more than just recognizing learned material. It requires referring to a theory, i.e., an organized set of explanations and predictions of facts (a set of relations), and deciding whether a given statement is or is not based on that theory.

To make this decision, the subject can operate at different levels. The grounds for the decision can be the cues provided by the acquisition context, for example: Were the pieces of information in the question presented in the same part of the course? This solution, which relies on memory, is risky but advantageous for persons who have not mastered the theory itself. In this case, the subject's uncertainty assessment is linked to his/her doubt about the items themselves. If no doubt is apparent, then there is strong confidence, even if the relation is unfounded. The probability of incorrect responses in the absence of doubt is thus high.

A subject can also evoke the concepts in the theory, and deduce from them the set of all possible relations between items. This will allow him or her to decide whether or not the relation proposed in the statement is compatible with the theory. In this case, the probability of a correct response increases. The confidence rating will be derived from the subject's doubt about the evoked consequences of the theory and/or a judgment of the compatibility between the consequences and the statement. With this kind of doubt, incorrect responses will occur, and hence, the confidence levels will be statistically closer to reality.

Realism in confidence assessment is indicative of a more fundamental realism: that of the understanding of the question posed, of the decision about the relevance of the involved information, of the mastery of the material in the problem to be solved. Lack of realism in confidence assessment in this case is indicative of the subject's inability to solve the problem at hand. Depending on the way the problem is stated, those who have not mastered the material may think they are solving it, in which case incorrect responses are accompanied by a high degree of confidence.

Note however that responses founded on indirect cues do not necessarily lead to overconfidence. This can be seen in the question presented below in example 2.

Example 2
Which of the three cities below is the farthest north?
  a - Prague
  b - Montréal
  c - Vladivostok
In this case, the subjects know they do not have enough information to determine the answer. But they can infer an answer, based on a cue like the climate, for instance. The outcome is as follows: if the question was constructed so that the climate cue would contradict the latitude cue, then the percentage of correct responses will be less than the chance level. But, in general, the confidence ratings are approximately equal to the chance level, which means that this is not really a case of overconfidence.

These examples, taken out of many similar ones, help us to understand why, even without a guessing strategy, many MCQ tests exhibit overconfidence. Insofar as non realism is often determined by a systematic organization of the judgmental activity, we agree with a lot of authors who criticize the use of confidence assessment as part of a scoring rule in summative evaluation, (for example, Sieber, 1974). However, in formative evaluation, the possibility of two kinds of errors, whether due to an inappropriate strategy or a partial knowledge only, should encourage teachers to use confidence assessment techniques. The results of the student's meta-cognitive activity enlighten the observer about the judgmental processes themselves.

1.3 How to reduce overconfidence (1): a conflict between critical judgment and confirmation bias

One possible way to reduce the overconfidence commonly observed with multiple-choice knowledge questionnaires is to encourage the subjects to do a thorough, critical analysis of the answers proposed. The experiment published by Koriat, Lichtenstein, and Fischhoff in 1980 addressed this question by having subjects search for the potential reasons why the proposed answers might be correct or incorrect. This search for multiple points of view turned out to be fruitful. When subjects had to state all the reasons why each choice might be correct or incorrect (non-selective instructions), or select the best reason why they might be incorrect (selective-contrastive instructions), a reduction in overconfidence was observed.

Note that in this type of experiment, overconfidence is treated as a whole, i.e. as an overall result characterizing the response patterns for a given questionnaire. A finer approach which accounts for the structure of the questions, i.e., the actual problem statements, would no doubt be preferable. Indeed, it is not uncommon in a questionnaire characterized globally by overconfidence, to find substantial variations across questions in this respect. The confidence ratings for certain questions, even some of the more difficult ones, can be quite realistic, while those of other questions can be indicative of underconfidence.

I replicated the Koriat, Lichtenstein, and Fischhoff (1980) study with a questionnaire about the material studied during freshman year psychology at a French university. The questionnaire contained multiple-choice questions for which the subjects were supposed to state which one of three choices was the correct answer. The questionnaire had two parts which differed in degree of difficulty, and which occurred in different orders across subjects. Confidence ratings were given in percentages, ranging from 40% (nearly chance level) to 100% ("totally sure"). As is customary in this type of study, the way the confidence rating scale was presented required the subjects to estimate the likelihood of response accuracy, and stressed confidence assessment realism.

The following two factors were crossed: (1) either the subjects commented about the selected answer or about the rejected answers, and (2) the comments were supposed to either support or criticize the subject's reponse.

The subjects wrote their comments in a special area to the right of each question. Subjects were given plenty of time to answer. They did not rate their confidence level on the scale provided below the comment area until they had completed the comments.

This gives us the following four experimental groups (table 3):

Table 3
Experimental groups in the comment experiment.

direction object	support decision	criticize decision
chosen answer	1. in favor of chosen answer	3. against chosen answer
rejected answers	2. against rejected answer	4. in favor of rejected answers

Subjects were assigned at random to one of the four groups, making 36 subjects in groups 1, 2, and 3, and 26 subjects in group 3.

In line with the results obtained by Koriat, Lichtenstein, and Fischhoff, a reduction in overconfidence could be predicted when the choice selected was to be criticized. However, I will not present the results for this prediction, since it did not turn out to be true. This experiment was interesting for a different reason. I will simply mention here that the general tendency was in fact overconfidence, and that this tendency was stronger in the difficult part of the questionnaire. But the manipulation of the comments did not have the expected effectiveness. The reason, we shall see, is probably the following: the effectiveness of having the subjects add comments depended on their ability to make such comments. In a general knowledge questionnaire, like Koriat, Lichtenstein, and Fischhoff's, we know that subjects make inferences based on uncertain cues and that they are easily confused because they are outside their usual domain of competency. They lack accurate landmarks to consider the various points of view. Instructions to critically examine the possible answers may make them aware of this situation. In contrast, when the questions deal with more specific and more recently learned material, as was the case in our experiment, the gap narrows between spontaneously elicited knowledge and that elicited by the instructions. The problem then becomes executing the activity required by the instructions.

The comments made by the subjects were coded by their object (chosen answer or rejected answers) and their direction (support or criticize the decision). Our first observation was the number of non-responses, i.e. the percentage of questions for which the subjects did not manage to come up with a comment despite the instructions. The distribution of these non-responses varied by the type of instructions and the difficulty of the question. Moreover, the object and direction of the comments did not always follow the instructions. This is shown in table 4.

Table 4
Comment experiment. Distribution of comments by question difficulty, and instructions stating direction and object.
%N: percentage of non-responses
%D: percentage of responses with the requested direction (support or criticize) out of the total number of questions answered.
%Dw: percentage of responses with the requested direction (support or criticize) out of the total number of questions.
%O: percentage of responses dealing with the requested object.
%O/d: percentage of responses dealing with the requested object among those with the requested direction.

| | support | | | | criticize | | | |
| | in favor of chosen answer | | against rejected answers | | against chosen answer | | in favor of rejected answers | |
	easy	diff.	easy	diff.	easy	diff.	easy	diff.
%N	9	5	23	20	41	35	47	53
%D	100	100	96	90	89	94	48	12
%Dw	91	95	74	72	53	62	25	6
%O	82	79	94	92	72	67	89	96
%O/d	82	79	94	91	68	66	100	86

The percentage of non-responses was extremely variable, ranging from 5% to 53%. This percentage depended essentially on whether the instructions said to argue in favor of the subject's response ("support") or against it ("criticize"). Among the former ("support" instructions), the percentages obtained opposed "arguing for the chosen answer" and "arguing against the rejected answers".

The distribution of the comments by direction provides us with another piece of information: the percentage of instructions which were in fact followed ranged from 100% (arguments in favor of the chosen answer) to 48% and 12%

(arguments in favor of the rejected answers, which were simply replaced by arguments against those answers).

Concerning the object percentages, when the instructions dealt with the selected answer, the correct object was more often discussed when the requested comment was positive ("support") than when it was negative ("criticize"). The %O and %O/d percentages indicated a tendency to substitute an argument supporting a rejected answer for a criticism of the selected answer. This happened approximately one out of three times.

The conclusion for this point is that simply instructing subjects to critically examine the proposed answers does not necessarily lead to satisfactory results. Doubt is a "luxury": it can only exist when the subject is actually capable of entertaining the various choices. This capacity is limited by the organization of his or her cognitive activity. Due to the availability of but a narrow set of fresh information, subject's activity clearly shows a strong *confirmation bias* (see Evans, 1989: "... human beings have a fundamental tendency to seek information consistent with their current beliefs, theories or hypotheses and to avoid the collection of potentially falsifying evidence"). As can be seen, the results obtained in this experiment were not determined by an artificial situation. Hence, we can hypothesize that such a confirmation bias would occur in most educational situations.

1.4 How to reduce overconfidence (2): providing an information about the structuring of possible events

Simply giving subjects a cue that allows them to reconstruct the set of all possible events (the possible choices) can be sufficient to create doubt and thus, to modify the confidence ratings.

This can be seen in example 3 and table 5 (taken from Fabre, 1980, experiment C). In this experiment, each question was given two different titles, a simple one and a multiple one. According to this factor, two different groups of students were compared.

Example 3
Normalization question (Q3)
Simple title: "Normalization"
Multiple title: "Normalization and quantile analysis"
In sampling by normalization of grades obtained for a variable x:
- the proportion of students per class varies
- the extreme classes include more values of x than the central classes
- there is usually an odd number of classes
- the extreme and central classes have the same number of students

Table 5
Effects of question title on the realism of confidence ratings (on a three points scale, where 1 means low confidence and 3 means high confidence): results obtained for questions 1, 3, and 5 where there was a strict correspondence between the two items in the multiple title and the two groups of statements in the questions (taken from Fabre, 1980, experiment C).

accuracy title	correct	incorrect
simple	2.37	2.31
multiple	2.45	2.16

Note that this effect has an impact on the confidence ratings but not on response accuracy. The information given the subjects was insufficient to improve their judgments, but it nevertheless allowed them to assess their knowledge.

1.6 Conclusions for this point

Previously published (Fabre, 1980) and new results thus allowed us to further our understanding of this issue by demonstrating:
    - the strong correlation between the judgment activity, which is reflected by the main responses, and its indicator, the meta-cognitive confidence rating, and
    - the sensitivity of the confidence ratings to certain difficulties encountered by subjects, and to certain aids offered to them, without causing any effects on the main responses.
    These two points are not contradictory. If we analyze the judgment task by looking at the structure of the questionnaire, we can see that variation in the degree of certainty is affected by the way the subject functions, and in particular, by *how well the material is understood and how the problem is encoded*. Confidence ratings are indeed a reflection of how much difficulty the subject experiences in choosing an answer. Depending on how a subject functions when faced with the problem, difficulties can be experienced at *various levels of relevance*, and can lead to certain abnormalities in the resulting confidence ratings.

2 The coding of expressions: Cognitive context does determine the differentiation between certainty and uncertainty

In the research tradition followed by Foley (1959) and Hogarth (1980), for example, which is based on the assigning of numerical values to verbal expressions of uncertainty, I conducted a series of experiments (thoroughly described in Fabre, 1991). A brief reminder of the main results obtained is relevant here, since these studies dealt specifically with the subject's processing of uncertainty.
    The basic principle behind this research project was as follows: a subject is presented with several verbal expressions representing ordered degrees of certainty, and is then asked to place each of these expressions on a numbered scale ranging from the least amount of certainty to the greatest amount of certainty. For example, in a previous study I presented to 143 subjects the following instructions:

Here are five common expressions that express levels of confidence that a given event will occur. You are requested to assign a number between 0 and 10 to each expression. Zero means total uncertainty about the fact that the event will occur. The number 10 means total certainty about the fact that it will occur.
    I suppose that the event will occur

It seems to me that the event will ocur
I think that the event will occur
I am sure that the event will occur
I affirm that the event will occur

Table 6 (taken from Fabre, 1993) gives the distributions of numerical responses for the five expressions. Theses distributions elicit not only a variability of the coding between the expressions, but also a variability between subjects, within each of the expressions.

Table 6

Frequencies of numerical coding for five expressions with a 0-10 scale

scale	I suppose	It seems	I think	I am sure	I affirm
0	3				
1	7	7	1		
2	17	19	2		1
3	26	17	4		
4	32	25	3	1	
5	2	39	31	1	2
6	16	23	39		2
7	11	9	30	3	
8	4	3	21	17	11
9	1	1	11	44	25
10			1	77	102
mean	*4.13*	*4.37*	*6.33*	*9.31*	*9.49*
standard-deviation	*1.84*	*1.72*	*1.57*	*0.97*	*1.12*

The numerical scale itself can be presented in different forms and used for different properties, but this is not the purpose of the research I am discussing here. The starting point for this series of studies was the following observation: when the lower limit of a numerical scale, i.e. the one which pertains to the least amount of certainty, is defined, it can represent one of two things: (1: UP scale) uncertainty - as in my initial research, or (2: NP scale) negative certainty, i.e. the certainty that an event will not occur - as in Foley (1959). See Foley's instructions:

Assuming the probability that an event X will occur can be assigned a value ranging from 1 (X will not occur) to 10 (X will occur), what values would you assign to the following statements: I am sure X will occur, (etc.).

With my UP scale, the numbers grow from Uncertainty to Positive certainty about the occurrence of the event. With Foley's NP scale, they grow from Negative to Positive certainty. Disregarding any experimental results, these two definitions of the lower limit of the rating scale appear very different, since "logically", uncertainty is situated somewhere between the extreme degrees of certainty. Yet the comparison of my first empirical results with Foley's convinced me that the distinction between these two definitions, which is linguistically clear, is not so obvious from a functional standpoint. Apparently, a statement like, "I suppose event X will happen" is coded in the same way, regardless of whether the lower limit of the scale is set at "uncertainty" or at "certainty that the event will not occur".

This led me to conclude that the linguistic distinction is not necessarily relevant. It only becomes functionally meaningful when it is associated with an adequate representation of the universe of all possible events. With probability of occurrence scales, which I would readily extend to probability of accuracy scales, there is no such thing as a clear representation of uncertainty, i.e. one which is distinct from the inverse of certainty, unless the subject is aware of the universe of all possible alternatives.

In order to draw this conclusion, let me outline the design used in the experiments. Two factors were crossed: (1) the definition of the lower limit of the scale used to numerically code the verbal expressions: uncertainty (UP scale) or negative certainty (NP scale), and (2) the problem referred to: an occurrence problem (for example, "I suppose event X will happen") or a choice problem (for example, "I suppose the car will turn right", for an intersection where there are only two possibilities, turn right and turn left).

The 447 subjects were randomly assigned to one of the four groups. Figure 2 shows the results pertaining to the functional interaction between the linguistic scale and the structure of the reference universe.

__Figure 2__
Effect of the definition of the scale's lower limit on the rating of three expressions, for the occurrence problem and the choice problem. (rating scale from 1 to 9)

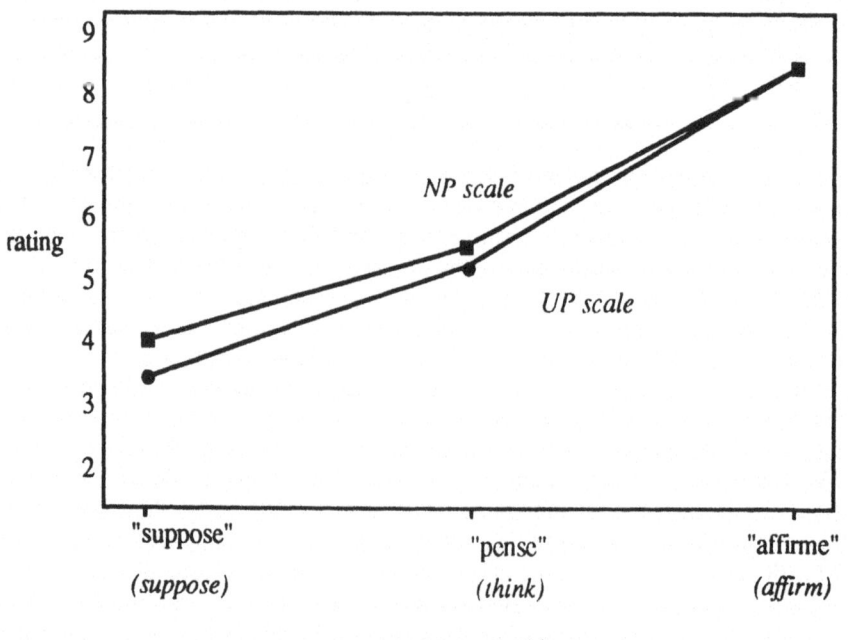

occurence problem

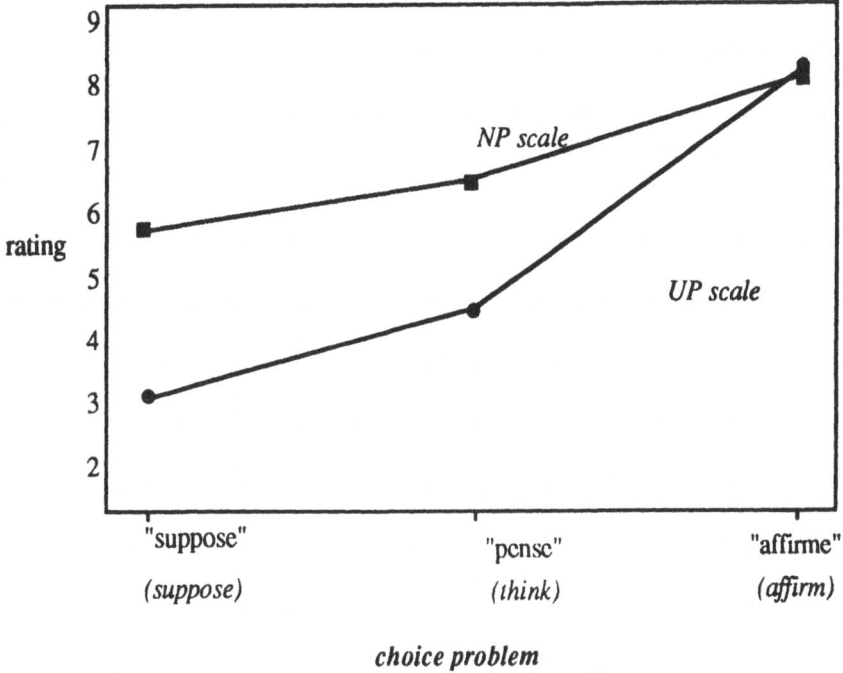

choice problem

We can see that in the case of a poorly structured, or unstructured, universe (occurrence problem), the difference between uncertainty and negative certainty was not very clear. With the UP scale, the occurrence problem gave rise to a higher mean rating than the choice problem. The opposite effect was obtained with the NP scale. The relationship between uncertainty and negative certainty in the occurrence problem can apparently be interpreted as reciprocal confusion.

Conclusion

Two related points should be considered in any attempt to understand the issue of realism in confidence assessment: the structure of the knowledge to be assessed, and the meaning of the degrees of confidence on the scale used to express that assessment. There are two distinct problems here, one dealing with information processing, and the other, with the judgment activity.

By relating our two approaches, we saw that the meaning of the scales used to express certainty is a problem that cannot be treated in isolation. This meaning is not fixed, and is affected by the cognitive context in which the information is processed. Consequently, we cannot pose the problem of the realism of certainty assessment without considering the structure of the questioned knowledge and the levels of questioning.

When low availability of the set of clearly delineated choices leads to confusion between certainty and uncertainty, we cannot expect subjects to be able to clearly affirm their state of doubt.

The reflection needed regarding the meaning of confidence rating scales has certainly been neglected by the success of the resolutely quantitative approach.

Indeed, when a numerical scale is used, when the instructions describe the numerical degrees on the scale as analogous to probabilities, and when the use of the scale is described in terms of its relation to a statistical norm, we can settle for a comparison between the broken line representing the empirical data and the diagonal representing realism.

Figure 3
Diagram of confidence realism. The diagonal represents realism (equality of degree of confidence and proportion of correct responses). The broken lines are the experimental data obtained for a rather difficult knowledge questionnaire and a rather easy knowledge quastionnaire. The location of the empirical data with respect to the diagonal reflects overconfidence or underconfidence.

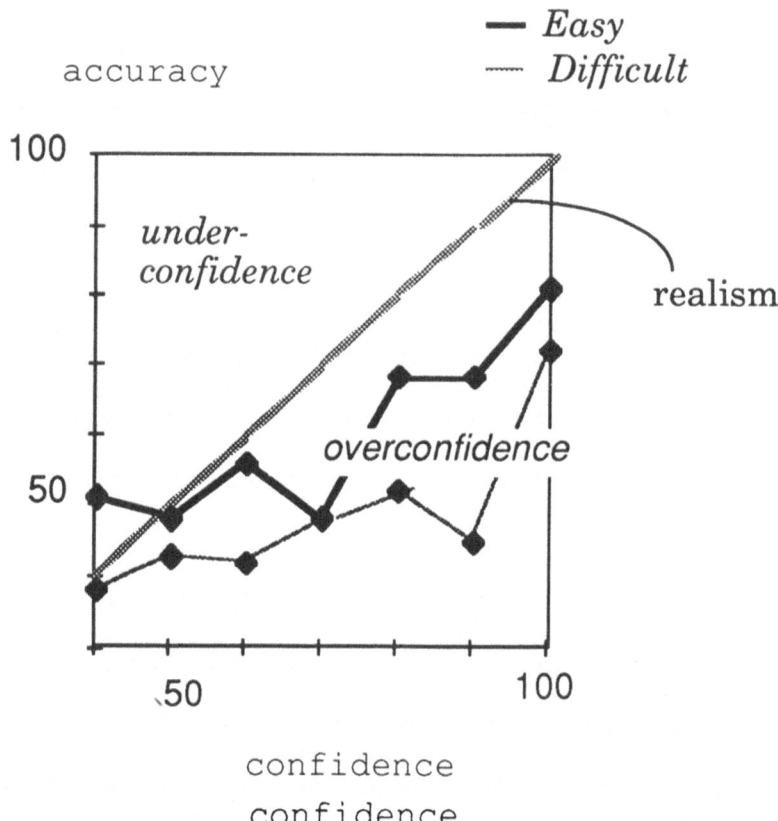

If the line is mainly under the diagonal, then we have overconfidence; if it is mainly above, we have underconfidence. But in both cases, the interesting point is that the overall slope of the broken lines is not as steep as that of the diagonal. Most likely, this means that the functional significance of at least one of the two limits of the certainty scale is not clear, in both cases. The reasons for this lack of clarity are linked to the subjects' representations of the problem context (the available information) and to the way they treat the questions asked.

It is likely that when subjects use these so-called quantitative scales, they implicitly code the degrees on the scale using common terminology, thus defining ordered classes which are operational for their own particular case. A finer application of the degrees on the scale probably corresponds to the intention to make within-class differentiations analogous to what is usually observed for rating scales in psychophysics and value judgments (Parducci, 1983, 1990).

Under this hypothesis, expressing doubt can have two different meanings:
- either it means we are aware of counterarguments which could turn out to be decisive,
- or it means that for reasons which remain confusing or unknown, the choice may turn out to be incorrect.

Perhaps it is a lot to ask of subjects to rate their level of confidence in the latter case. And by confounding these two cases, we are most certainly making an error of analysis which we could avoid by using a more cognitive approach.

References

Evans, J. St. B. T. (1989). *Bias in human reasoning: causes and consequences*. Hove and London: Lawrence Erlbaum.

Fabre, J.-M. & Noizet, G. (1977). Jugement du vrai et du faux sur des énoncés affirmatifs et négatifs. In *Psychologie expérimentale et comparée: Hommage à Paul Fraisse* (pp. 411-426). Paris: Presses Universitaires de France.

Fabre, J.-M. (1980). *Jugement et certitude, Recherche sur l'évaluation des connaissances*. Berne: Peter Lang, Collection Exploration.

Fabre, J.-M. (1991). The expression of uncertainty: two contextual effects. *European Journal of Cognitive Psychology, 3*, 399-412.

Fabre, J.-M. (1993), *Contexte et jugement , de la psychophysique à la responsabilité*, Lille: Presses Universitaires de Lille.

Foley, P.J. (1959). The expression of certainty. *American Journal of Psychology, 53*, 614-615.

Hogarth, R.M. (1980). *Judgment and choice, The psychology of decision*. New-York: Wiley.

Koriat, A., Lichtenstein, S. & Fischhoff, B. (1980). Reasons for confidence. *Journal of Experimental Psychology: Human Learning and Memory, 6*, 107-118.

Lichtenstein, S., Fischhoff, B. & Phillips, L.D. (1982). Calibration of probabilities: The state of art to 1980. In D. Kahneman, P. Slovic & A. Tversky (Eds.), *Judgment under uncertainty: Heuristics and biases* (pp. 306-334). New York: Cambridge University Press.

McClelland, A.G.R., Coulson, A.S. & Icke, S.E., (1990). Bias in meta-memory performance and its implications for models of memory structure. In J.-P. Caverni, J.-M. Fabre & M. Gonzalez (eds), *Cognitive Biases* (pp. 511-519). Amsterdam: North-Holland.

Noizet, G. & Fabre, J.-M. (1975). Etude docimologique des questionnaires à choix multiple. *Scientia Pædagogica Experimentalis, 12*, 38-62.

Parducci, A. (1983). Category ratings and the relational character of judgment. In H.-G. Geissler, H.F.J.M. Buffart, E.L.J. Leeuwenberg & V. Sarris (Eds.), *Modern issues in perception* (pp. 262-282). Amsterdam: North-Holland.

Parducci, A. (1990). Response bias and contextual effects: when biased? In J.-P. Caverni, J.-M. Fabre & M. Gonzalez (eds), *Cognitive Biases* (pp. 208-219). Amsterdam: North-Holland.

Sieber, J.E. (1974). Effects of decision importance on ability to generate warranted subjective uncertainty. *Journal of Personality and Social Psychology, 30*, 688-694.

Validity, Reliability, and Acuity of Self-Assessment in Educational Testing

Dieudonné Leclercq

Université de Liège, Service de Technologie de l'Education, Batiment 32, Sart Tilman, 4000 Liège 1, Belgium, Tel. (32 41) 56 20 72, Fax (32 41) 56 29 44, e-mail: U017801 at BLIULG11

Abstract: When teachers use confidence marking, they should be aware that confidence estimation and confidence expression are influenced by a series of factors. Some of them have been studied in detail, such as the general human capacity to estimate one's knowledge (how far can people be sensitive, reliable and valid in appreciating their uncertainty).

This paper indicates how some of these factors have been studied, the results and the implications for designing test instructions, proper scoring rules and indices of the quality of self assessment.

Keywords: Subjective probabilities, self assessment, validity, reliability, sensitivity, confidence marking

The expression of a student's degree of confidence depends upon two series of factors: factors affecting the confidence *estimation* and factors affecting the confidence *expression*.

A. Factors affecting the confidence estimation

1. The student's ability

The student's ability should be validly and reliably reflected by the confidence degree.

The degree of ability could influence the student's judgment in two respects:

a) Students with perfect (100% correct) ability are likely to underestimate (it is impossible to overestimate). The reverse is true for student with 0 ability level.

b) Sensitivity is likely to be more subtle in the extremes (low probabilities and high probabilities) of the scale, as Edwards (1971) advocated by noticing that human estimators use ratio scales. Therefore, they can distinguish 95% (1 chance out of 20 to be wrong) from 99 % (1 chance out of 100 to be wrong) but not 50 % (1 chance out of 2) from 54 % (1 chance out of 2,17). This fits with what we know about psychophysics: sensitivity (acuity) is not in a linear, but in a logarithmic progression with the probability scale.

2. The human average capacity of self-estimating

a) The human sensitivity (or acuity or granularity).
We know we are unable to make a difference between two confidence degrees as close as 37% and 38%. But how far can an average person distinguish on a probability scale? How many different degrees? Is Miller's (1956) "magical number seven plus or minus two" applicable in this domain?
b) The human realism (the validity problem): how far is it reasonable to expect an average person to self-estimate? Does it vary from content to content? How many tests do we need to assess the validity of self-estimation ? Does it depend upon the number of degrees?
c) The human stability in time (the reliability problem): does it depend upon the numbers of degrees?
   Obviously, those three concerns are interrelated and we shall provide data that answer several questions in the same time.

3. The personal sensitivity, reliability and validity

In this domain as in others in psychology, it is likely that interindividual differences occur, at least with untrained persons.

4. The content on which estimations are made

The content on which estimations are made could produce intrapersonal variations (a person may be more realistic in some domains than in others). As an answer to one of our questionnaires, testees noted that "they would express their probability differently if they were risking their life". There also exist domains where a bit of irrealism helps overcome difficulties of life, such as the 40-year old man saying: "I have already lived one third of my lifetime!"

5. The degree of familiarity and of training with the specific procedures used

As shown in Section J, self estimation can be inproved by practice, especially if the learners are exposed to the consequences (payoffs) of their actions, in an operant conditioning way.

B. Factors affecting the confidence expression

There can be a difference between what a person believes and what he/she expresses (says or writes) for a series of reasons.

6. The qualitative aspect of instructions

The qualitative aspect includes such effects as that of using different types of scales, for example,
   - An ordinal scale (weakly sure, fairly sure, strongly sure), i.e. the more widely used instructions (unfortunately!).

- A continuous confidence marking system (De Finetti, 1965a) where the student expresses his/her confidence with any precision he/she likes (e.g. 0.3 as well as 0.318), with the computer applying a scoring function (see details in Leclercq, 1983).

- A 10-stars system (Michael, 1968) where the student has to distribute the ten stars (each representing 10 % confidence) among the alternatives. From his experimental data, Michael concluded that a classicaly scored test should be 1.7 times longer to reach the reliability obtained with a system of ten stars.

- A 5-stars system (De Finetti, 1965a)

- Locating the answer within a (triangular) space (De Finetti, 1965; Bruno, 1993)

- A choice between some precoded zones of the probability scale (Leclercq, 1983 and Leclercq et al., 1993).

Although they are profusely used, ordinal scales lead to almost uninterpretable data. We all should consider "admissible probability measurement procedures" (Shuford et al., 1966) for assessing partial knowledge.

7. The consequences often represented by the scale of tariffs

Inappropriate scales could elicit lies to optimize the total amount of points, i.e. the final score. The scales of tariffs should, consequently, have been constructed carefully, according to decision theory (Van Naerssen et al., 1965) and be kept clear for the student so that he can compute its own score himself (Raiffa, 1970; Leclercq, 1983, p. 207).

8. Personal attitude towards risks

Atkinson (1971) has described persons motivated by the "hope of success" and others by the "fear of failure". Coombs has described (in his "unfolding theory") preferences for amounts of risks and of probabilities. The problem of conservation of expected utility can be formulated in the classical piagetian way: some are non conservants; for instance they are overinfluenced by the consequences, others by the probabilities (they do not apply the compensation principle).

9. Need for points

Students having few points and wanting to maximize their gains in order to reach the pass/fail threshold could be inclined to adopt a more cautious behaviour than people that know that their total score will anyway be beyond the passing score. Van Naerssen et al. (1966) have studied this problem, concluding that there is no reason for rejecting the assumption of linearity in the students' motivation of maximizing their total score.

C. The general picture of factors affecting confidence degrees

Those considerations lead to the following schema:

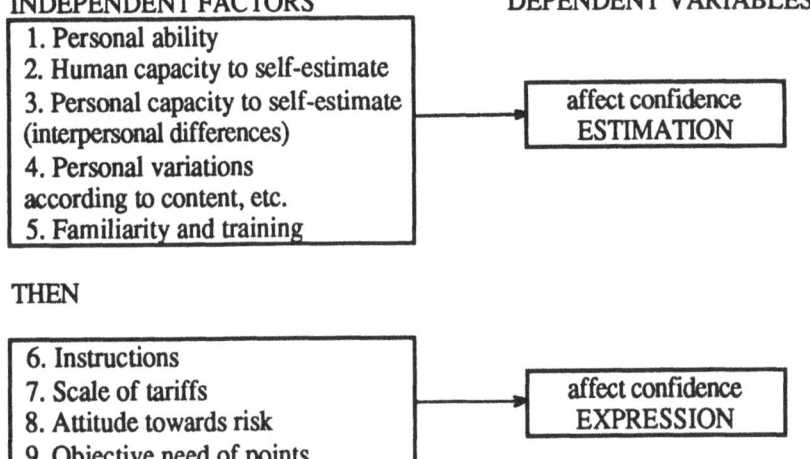

INDEPENDENT FACTORS   DEPENDENT VARIABLES

1. Personal ability
2. Human capacity to self-estimate
3. Personal capacity to self-estimate
(interpersonal differences)
4. Personal variations
according to content, etc.
5. Familiarity and training

affect confidence
ESTIMATION

THEN

6. Instructions
7. Scale of tariffs
8. Attitude towards risk
9. Objective need of points

affect confidence
EXPRESSION

Factors 3, 4 and 5 are sometimes refered as   personality factors and there is a reluctancy in "melting" them with the assessment of knowledge. As will be shown in sections I and K hereafter, individual realism can be "computed" and its weight in the score fixed at will.

Teachers are interested in eliciting the best student's estimate and having it expressed without bias.

The following lines will provide data on several of the components of the first part of the schema hereover. Other publications (Leclercq, 1983, 1990, 1993) have been dedicated to the second series (6, 7, 8, 9) of factors.

D. An experimental approach

In the following experiment, three problems (validity, stability and acuity) are studied in the same experimental design based on a "confidence guessing game" (CGGame).

1. The Sentence Confidence Guessing Game (SCGGame)

The confidence guessing game presented here is directly inspired by Shannon's guessing game (1951) in which the subject has to predict successively each letter of an English text. There are only 27 possible answers (each of the 26 letters, plus the "blank" for the spaces, points, etc.). In Shannon's method, when an answer is wrong, the subject has to guess other letters until he finds the correct one. The experimental data are presented by indicating below each letter the number of trials needed until the correct answer was found.

We slightly changed this game by allowing the student to provide only one guess (a letter) and by asking him/her to accompany his/her answer with a confidence degree. The player is provided with the correct answer after each trial. We will refer to this as the Letter Confidence Guessing Game (LCGG).

For the validity/reliability/sensitivity experiment, we used a "Sentence" Confidence Guessing Game (SCGG), in which a long text (about 3 pages) is chosen from a book. The odd lines of the text are printed and truncated whereas the even ones are not. The subjects are asked to predict the first letter of the truncation.

Cutoff points have been chosen in order to obtain items of various difficulties (ideally, a rectangular distribution with an average mean of 0.50). Figure 1 shows the distribution of the facility indexes for the 100 questions in the experiment.

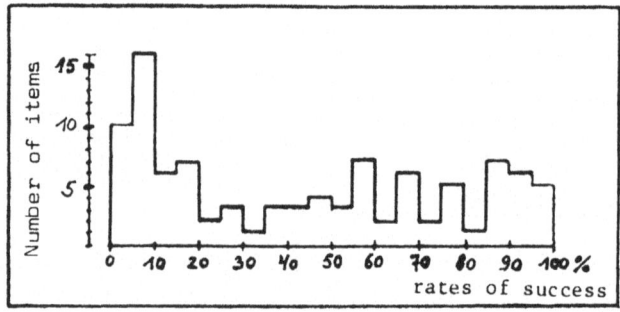

Figure 1

2. The three confidence degrees scales

Figure 2 presents an example of such an item. Here, the correct answer is L, since the truncated text is "The magical number seven plus or minus two." The subjects were requested to write the next letter (here the letter L) and to "circle" a subjective probability on each of the three scales.

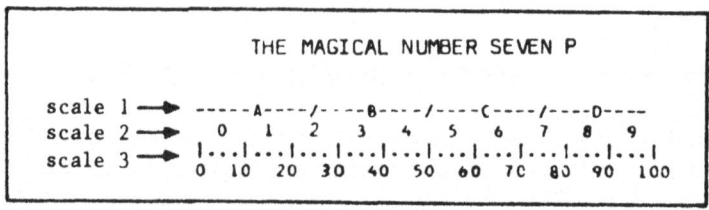

Figure 2

3. The tariffs

The subjects were told that "points would be given in such a way that, if they want to maximize their total score, they should not bias their subjective estimate" (i.e. they should tell the truth). The table of wins and losses (tariffs) for given probabilities, as well as their plotting, were presented to the students (see Figure 3). The maximal score is +50 (TC for confidence degree 100 on scale 3 of Figure 2) whereas the minimal score is –100 (TI for confidence degree 100 on scale 3).

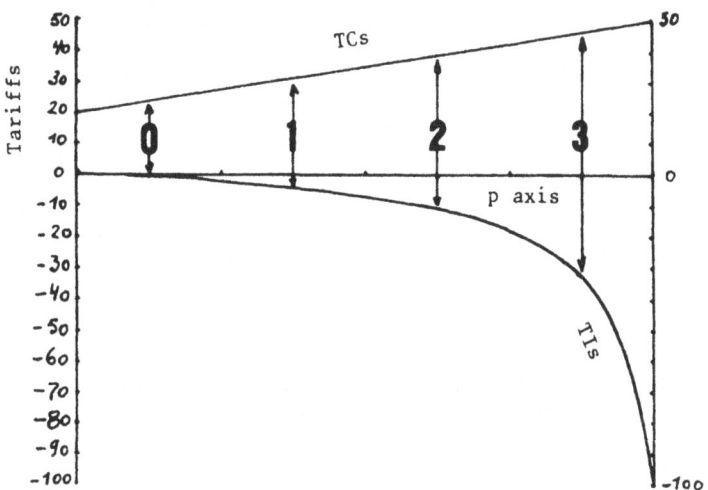

Figure 3

4. The experimental setting

The experiment was conducted in three steps:
a) The "guessing game" and the scoring rules were explained to about 300 high school teachers. A dry run was conducted with 5 items; each participant received the correct answer and his score a few days after (by mail). The results appeared on computer listings, and comments were given such as each participant's rank, or his/her overall tendency to overestimate or to underestimate.
b) In the experiment itself (test), subjects were requested to answer 100 items and to assign to each answer a subjective probability (SP) of correctness. SP had to be expressed on three different scales:
  - Scale A: 4 possible confidence degrees (25% each);
  - Scale B: 10 possible confidence degrees (10% each);
  - Scale C: 40 possible confidence degrees (2.5% each).
c) One month later (retest), subjects received the same questions and their answer (they were not allowed to change the answers), but did not receive their previous SPs. Subjects were requested to give their SPs again. Furthermore, on retest, subjects were invited to describe the way in which they chose their degrees of confidence.

5. The general results

The whole test-retest procedure was completed by 121 subjects. Figure 4 shows that the distribution of the 121 simple scores (SST = number of correct letters) is close to a normal distribution, with extremes being 39 and 66, with a mean of 56.

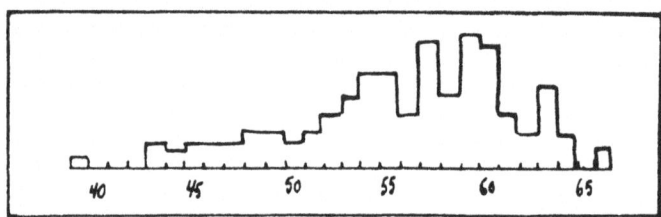

Figure 4

For 78 subjects, confidence scores (SCT: score computed with the confidence tariffs) were better on the first test than on the retest; for 42 subjects, the contrary occurred. Only one subject had the same score for both testings. The loss in efficiency on the retest could possibly be attributed to boredom having to answer 100 questions yet again. Therefore, the subjects could have estimated their confidence degrees less carefully, dedicating less time to this process.

E. Results concerning stability

1. The basic data

For each subject, a computer program prints:
- The scatter plot of the 100 confidence degrees used on the test (t1) and on the retest (t2);
- The two means (M1 and M2) and standard deviations (SD1 and SD2) of those degrees;
- The Bravais Pearson correlation coefficient (R) between the t1 and the t2 degrees;
- The histogram of the 100 degrees of confidence on the test;
- The histogram of the 100 degrees of confidence on the retest;
An example of the printout is given in Figure 5.

2. The results

Here are two examples (for subjects 104 and 94) of the two histograms (before and after test and retest) with a rather good replication pattern (a correlation of 0.80).

```
YOUR CODE = 104                          YOUR CODE = 94
(BEFORE) 40 DEGREES                      (BEFORE) 40 DEGREES
 3 *                                      3 ****************************
 7 *                                      7 *************
11 ***                                   11 ****************
15 ***                                   15 ****
19 **************                        19 **************
23 ****                                  23 ******
27 ***********                           27 **
31 ********************                  31 *********
35 *****************                     35 ***
39 *************************             39 ******
YOUR CODE = 104                          YOUR CODE = 94
(AFTER) 40 DEGREES                       (AFTER) 40 DEGREES
 3 *                                      3 ****************************
 7 *                                      7 *************
11 *                                     11 ****************
15 ***                                   15 *******
19 ****                                  19 **********
23 ***********                           23 ****
27 **************                        27 ******
31 ******************                    31 **********
35 ***************                       35 ****
39 *************************             39 *
```

Figure 5

For the 121 subjects, the Median correlation observed is 0.56 (not so bad for a first trial by untrained subjects).

```
                                      X
                        X             X                  X
                       XX             X    X   X X      X   X
                       XX             XXX XX XXX X       X   X
                       XX             XXXX XXXXXX X X   XXX X
            X   X X XXXXXXX           XXXXXXXXXXXXXXX X    XXXXX
 X    XX   X   XXXXXXXXXXXXXX         XXXXXXXXXXXXXXXXXXXXXXXXX    XX

 . 20        . 30        . 40        . 50        . 60     . 70        . 80
                                            Median
```

F. Results concerning acuity

1. Spontaneous uses of the 40 degrees

Whereas 40 degrees could be used, only 11 out of them have been systematically used, may be because the preprinted formula sheet was not attractive for "intermediate degrees".

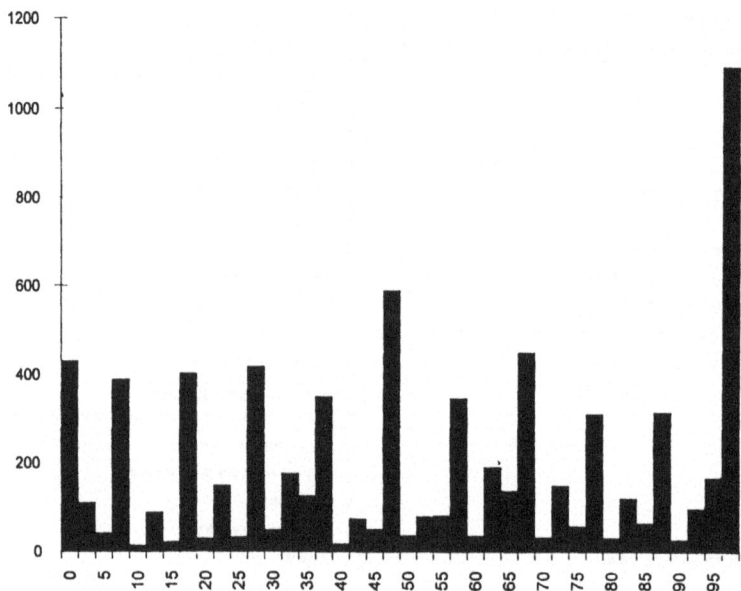

Figure 6

2. The Replication Histograms (or profiles)

A "replication histogram" has been built for each degree of confidence of each scale (4 histograms for scale A, 10 histograms for scale B, 40 histograms for scale C).

The replication histogram of a given degree (say X) is established according to the following principles (see Figure 6).

- The various degrees are placed on the horizontal line.

- The height of each rectangle expresses the number of times (here the percentages) that each degree (Z) has been used on the retest at the very place where degree X was used in the test. Actually, when degree X has been used in a test, degrees close to degree X are used in the other test (and degree X should be the most used of all of them).

In order to make the graphs clear, the tops of the histogram rectangles have been joined. Only these "profiles" are presented. The replication profiles computed from the 40 most realistic subjects for the 4 degrees of the rawest scale (A) are presented in Figure 7 (confidence degrees 0 and 3) and Figure 8 (confidence degrees 1 and 2).

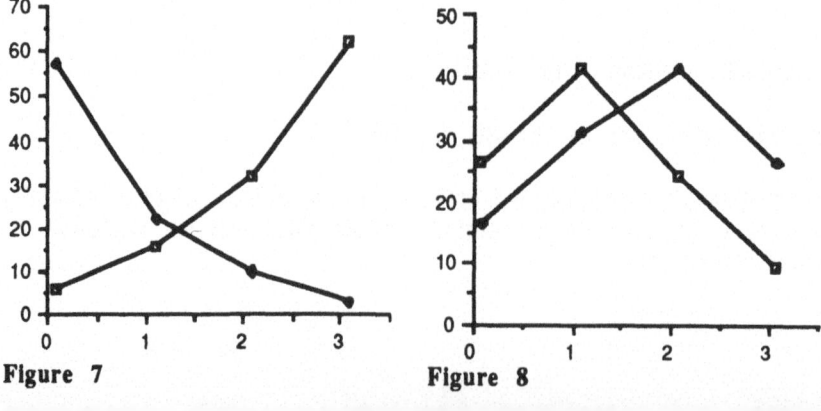

Figure 7   **Figure 8**

As can be seen, the top of the replication profile for degree 0 is 0, it is 1 for degree 1, etc. Our adult untrained subjects seem to have had no problem in dealing with 4 degrees (scale A) and it is likely that they can handle more sophisticated scales.

Figure 9 presents the 10 replication profiles for the B scale for 34 persons. It appears that the mode for X is X except for confidence degrees 7 (mode is 8), and 3 (mode is 4), and that for degrees 2, 4, 5 and 6 the peak is not neat!

This "overlap of some degrees" seems to indicate that 10 degrees is too much, either for untrained individuals or for this kind of work.

It seems also obvious that replication is better at the extreemes (low side or top side of the scale) than in the middle of the probability scale. Edwards (1967) has provided a possible explanation about that: "Human beings think not in probabilities but in ratios or odds. For instance, 99% is 1 chance out of 100 of being incorrect, whereas 95% is 1 chance out of 20 and 90 % is 1/10. Those ratios are quite different from each other. Whereas in the center of the scale 50% (1/2) is too close to 46 % (1/2.17).

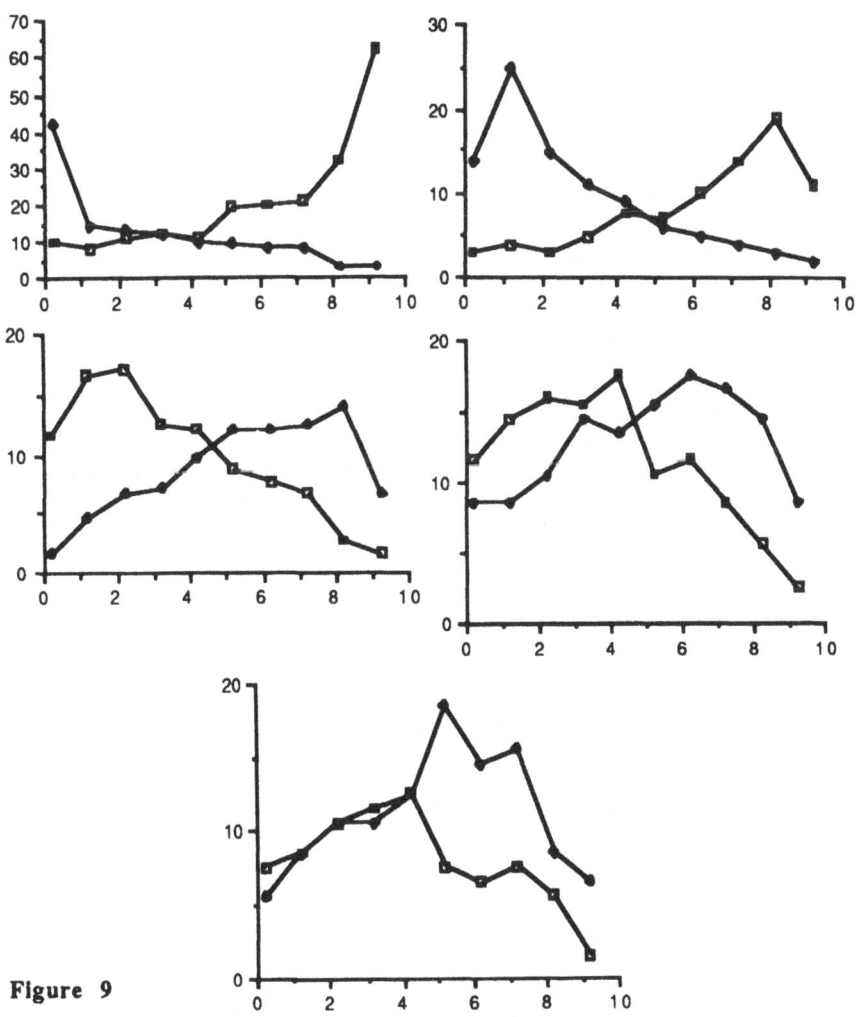

Figure 9

3. Subjects' introspection

Various questions have been asked to the subjects in order to know how they chose a given degree of confidence.

a) *Which sequence do you follow when you use the scales to indicate your confidence degree?* (4, 10, 40) or (40, 10, 4) or (10, 4, 40), etc. ?
Here are the answers in decreasing order of importance (number of observed cases):

1.	4, 10, 40	(56)
2.	40, 10, 4	(18)
3.	10, 40, 4	(15)
4.	10, 4, 40	(11)
5.	40, 4, 10	(10)
6.	4, 10 or 40	(9)
7.	40, 10 or 4	(7)
8.	10, 40 or 4	(6)
9	4, 40, 10	(3)

Starting with the 4 degrees scale (top down method) is used by 68 persons (50%).
Starting with the 40 degrees scale (bottom up method) is used by 35 persons (26%).
Starting with the 10 degrees scale (intermediate) is used by 32 persons (24%).

b) *How do you come up with a specific degree?*

Here is a selected series of answers:
   - When I am perfectly confident, I start from 100% and when I am not confident at all, I start from 0%.
   - I point a given place on the line (from 0 to 100) with no concern for any numerical value.
   - For me there are three situations : sure, doubt (50%), not sure at all (0 to 10%). My "sure" responses are subsequently divided into "perfectly sure, i.e. 100%", "very likely, i.e. 95%", and "likely, i.e. 80%".
   - I first gave my confidence degrees for the answers I was perfectly confident (100%), then to the rest of the items.

c) *On what does the choice of a confidence degree depend?*

The 135 interviewed persons answer as follows:
   28%: mainly on my subjective probability AND subsequently on risk.
   26%: only on my subjective probability.
   19%: on my subjective probability and on risk, equally.
   18%: on risk only.
    9%: mainly on risk, and subsequently on my subjective probability.

The popularity of the three first propositions (73 %) is a good indicator for the validity of the whole procedure. Some subjects noted that their strategy depends upon the situation (circumstances and objectives): if they were to risk their life, their strategy might change.

d) *What is your ideal number of degrees on the probability scale?*

10 degrees:	41
20 degrees:	13
40 degrees:	8
100 degrees:	4
10 to 20 degrees:	3
4 degrees:	3
50 degrees:	2

Others (5d., 6d., 7d., 8 d., 7 to 10 d., 4 to 10 d., etc.)

Some subjects that have chosen 10 degrees as the ideal scale added interesting comments:

-With the possibility of rating + and - for each degree (for example 7+ and 7-), this comes close to the 20 degrees scale.

-With the possibility of using some special intermediate values as 25%, 75%, 95%.

A subject explained that he has chosen the 20 degrees scale because he used it in scholar setting.

Other interesting comments were made:

- My ideal scale is 10 degrees because there were 100 questions; with only 5 questions, 40 degrees would have been perfect.

- Why not a continuous confidence marking (for example 39%)?

- This game can be learned and subject could improve their ability.

- I would prefer to give several answers and give to each a probability (this teacher was a linguist).

In general, subjects pointed out the importance of the number and the kind of items, of the situation (real consequence or not) of the subject (if he is "words minded" or not) and of the content of the text.

4. Conclusions for acuity

Conceiving a perfect test-retest situation is a hard job. To insure a same level of uncertainty on the test and on the retest, subjects must not reformulate the hypothesis about the correct answer and, consequently, change the probabilities.

We have not succeeded in building a situation where subjective probabilities are expressed given fixed hypothesis. Such a conditional game should be developed.

From our Confidence Guessing Game (CGG), it has been possible to observe reasonable test-retest stability.

On the resolution side, objective data as well as subjects' opinion show that for this kind of game, for this kind of (untrained) graduate adults, the maximal number of degrees was between 6 and 8. This result corroborates G.A. Miller's opinion that our spontaneous resolution in perceptual domains is "the magical number seven, plus or minus two".

Moreover, we can formulate the hypothesis that our "sensibility" (or resolution or acuity) is better in some portions of the probability axes (the extremes) justifying W. Edwards' procedure (1967) of using odds (that have logarithmic properties).

All those observations are of interest for conceiving an optimal scale to be used in school settings (Leclercq et al., 1992). There must only be a few degrees for practical reasons, but how many exactly and where on the axis of probabilities?

The present study can inspire new experiments to answer more precisely those questions.

G. Results concerning validity

1. With the 10 degrees scale

The calibration curve of 20 selected persons highly realistic (see Figure 10) shows the ambiguity between degrees 3 and 4 and between degrees 5 and 6.

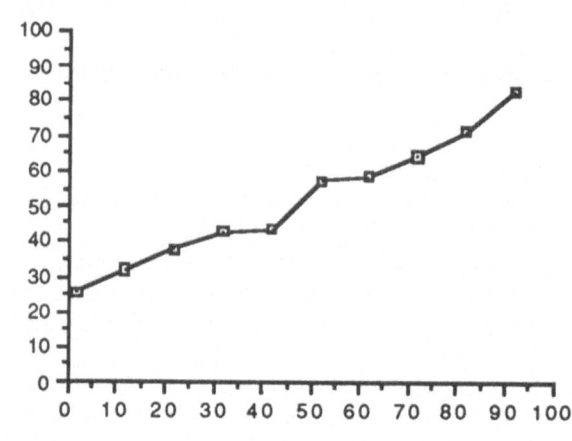

Figure 10

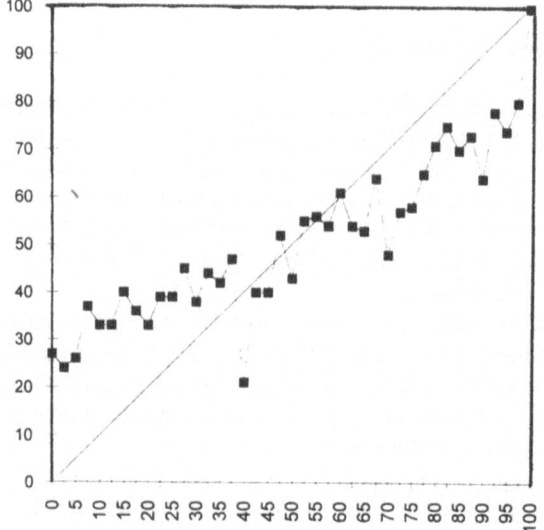

Figure 11

Things happen as if people had their best acuity (accuracy) at the extremes of the scale, since degree 0, 1, 8 and 9 are never confusing. These results indicated that less than 10 degrees should have been used, since our subjects had difficulties in handling 10 degrees.

2. With 40 degrees scale

It becomes impossible to build replication profiles, because some degrees have not enough observations. Figure 11 presents the general calibration-curve, where reliable values (computed on sufficient data) are represented by dark dots.

This calibration curve reinforces the impression (already given by the "acuity results") that adults cannot, obviously self-estimate validly with 10 degrees, but probably with three or four less, coming close to Miller's famous "magical number seven".

3. Ceiling and floor effects

A highly competent person (who succeeds in almost 100% of the occasions) can hardly overestimate! The same for a very incompetent one (succeeding close to 0% of the trials): he/she can hardly underestimate. Therefore, (normal or small) underestimation is to be expected for the former and overestimation for the latter. This explains that the slope of the calibration curve is lower than 1 (the diagonal line).

H. Formulas to compute the indices

We have developed the following formulas, based on previous research (Brier (1950), Adams & Adams (1961), Oskamp (1962), Murphy (1972, 73, 74) and Lichtenstein et al. (1977)), we have described elsewhere (Leclercq, 1983):

Acuity (or subtlety): The tendency to diversify one's judgments.
Formula: Standard deviation of the various rates of successes of the various degrees used.

Centration: Difference between the rate of (objective) success and the average confidence degree (result is either positive, negative or zero - perfection).
Formula: Difference between Average central values of confidence degrees used (or average confidence) and Rate of correct answers (percentage).

Coherence: The tendency of the successive rates of correct answers to fit a straight line (even if it is not the diagonal).
Formula: Correlation (Bravais-Pearson) between rates of successes and central values of the successive zones (when the degree has been used).

Realism: The tendency of the rates of correct answers to fit the diagonal.

Formula:

$$ERR = \sqrt{\frac{\sum (RS_i - CV_i)^2 \cdot NU_i}{NA}}$$

where   RSi = Rate of Success of confidence degree i.
        CVi = Central Value of confidence degree i.
        NUi = Number of uses of confidence degree i.
        NA = Number of answers.
        ERR = a quadratic expression of error of realism.

$$\boxed{\text{REALISM} = 1 - (5\ \text{ERR})}$$

I. The need for norms

ACUITY is ...	if it is	% of 311 students
IDEAL	= 50 %	0 %
EXCELLENT	superior to 25 %	14 %
GOOD	between 23 % and 24 %	12 %
SATISFACTORY	between 20 % and 22 %	30 %
WEAK	between 16 % and 19 %	27 %
INSUFFICIANT	less than 16 %	15 %
MINIMAL	= 0 %	2 %

CENTRATION is ...	if the difference is	% of 311 students
IDEAL	= 0 %	1 %
EXCELLENT	less than 3 %	18 %
GOOD	from 3 to 6, 99 %	19 %
SATISFACTORY	from 7 to 10, 99 %	19 %
WEAK	from 11 to 15, 99 %	17 %
INSUFFICIANT	16 % and more	26 %
MINIMAL	97.5 %	0 %

COHERENCE is ...	if the correlation is	% of 311 students
IDEAL	= 1	0 %
EXCELLENT	superior to 0.97	13 %
GOOD	between 0.93 and 0.96	18 %
SATISFACTORY	between 0.85 and 0.92	28 %
WEAK	between 0.75 and 0.84	18 %
INSUFFICIANT	less than 0.75	23 %
MINIMAL	- 1	0 %

REALISM is ...	if the index is	% of 311 students
IDEAL	= 1	1 %
EXCELLENT	superior to 0.95	15 %
GOOD	between 0.91 and 0.94	18 %
SATISFACTORY	between 0.84 and 0.90	22 %
WEAK	between 0.70 and 0.83	22 %
INSUFFICIANT	less than 0.70	22 %
MINIMAL	- 3.75	

Those values are just provisional references. Concurrent and predictive validity measures (with achievement and grades) are currently computed in experimental settings. Simple correlations could be misleading since overestimation (in uncompetent persons) may be worse than underestimation (in competent persons) whereas their realism index may have the same values.

In addition, intrapersonal variations of realism due to different contents needs also to be explored, when familiarity (and subsequent quality of realism) with the procedure has been reached.

J. Can realism be improved by practice?

With the help of F. Lambert (1992), we have studied the evolution of these 4 indices for (86) students involved in four settings:

1. A home answered test (March 92). It is supposed to be easy since no constraint of time exist and the students could discuss and cooperate with each other to answer the test.

2. A first computer test (end of April 1992). It is supposed to be the most difficult since each student has to answer alone, with time pressure.

3. A second computer test (May 1992). Easier than the previous one since familiarity with the computer testing situation has developed.

4. The last test (beginning of June 1992), paper administered. The easiest since familiarisation to the whole procedure has occurred, there is no time pressure and the students have an overview of all the questions (since it is on paper).

Here is the evolution of the average values of the 4 indices:

	Acuity	Centration	Coherence	Realism
Max	35	0	1	1
1.Home(March)	18.8	13.4	0.38	0.56
2.Comp.(April)	22.4	17.5	0.31	0.40
3.Comp.(May)	21.5	16.5	0.45	0.52
4.Paper (June)	21.5	10.6	0.78	0.80

Centration, Coherence and Realism have improved continuously during the three "real" tests (i.e. really valued for the final score), i.e. tests 2, 3 and 4.

K. Conclusions

Conceptual and technical tools exist to assess students' acuity, centration, coherence and realism of self-estimation of competency. Norms are developing. This behaviour can be trained. It is worth being explored further.

Just like Bruno De Finetti (1965b), we believe that
"Partial information exists. To detect it is necessary and feasible" (p. 109)
and that
"It is only subjective probability that can give an objective meaning to every response and scoring method" (p. 111).

This meaning has not always been well interpreted. For instance, scoring has been confused with measurement. The previous issues resulted in a first wave research proved to be a dead end in the late seventies. Nowadays, a new wave is developing rapidly. Prototypes of this movement are Shuford's, Hunt's and Bruno's work as well as the TASTE Approach, pieces of which can be found in Leclercq & Bruno (1993).

References

Adams, J.K. & Adams P.A. (1961). Realism of confidence judgments, Psychological Review 68, 33-45

Attneave, F. (1959). Application of information theory to psychology. New York: Holt, Rinehart and Winston.

Brown, T.A. & Shuford, E.H. (1973). Quantifying uncertainty into numerical probabilities for the reporting of intelligence (Report R-1185-ARPA), Santa Monica, Cal.: Rand Corporation.

Bruno, J. (1993), Using testing to provide feedback to support instruction: a reexamination of the role of assessment in educational organizations. In: D. Leclercq, J. Bruno (eds.): Item banking: interactive testing and self-assessment. NATO ASI Series F, Vol. 112. Berlin: Springer-Verlag (this volume).

Coombs, C.H. (1950). Psychological scaling without a unit of measurement. Psychological Review 57, 145-158.

Coombs, C.H., Dawes, R.M., Tversky, A. (1970). Mathematical psychology. Englewood Cliffs NJ: Prentice Hall.

De Finetti, B., (1965a). La décision et les probabilités, Revue des Mathématiques pures et appliquées, Bucarest, 405-413.

De Finetti, B. (1965b), Methods for discriminating levels of partial knowledge concerning a test item, British Journal of Mathematical and Statistical Psychology 18, 87-123.

Edwards, A.L. (1967). Statistical methods. New York: Hold, Rinehart and Winston, 2nd edition.

Edwards, W., (1967), Probabilistic information processing by men and man-machine systems, in La simulation du comportement humain, Paris, Dunod, p. 187.

Hunt, Darwin P. (1993), Human self assessment - theory and application to learning and testing. In: D. Leclercq, J. Bruno (eds.): Item banking: interactive testing and self-assessment. NATO ASI Series F, Vol. 112. Berlin: Springer-Verlag (this volume).

Leclercq, D. (1975). L'évaluation subjective de la probabilité d'exactitude des réponses en situation pédagogique. Thèse de doctorat en Sciences de l'Education, Université de Liège, Institut de Psychologie et des Sciences de l'Education.

Leclercq, D. (1983), Confidence marking, its use in testing. In: Postlethwaite, Choppin (eds.) Evaluation in Education, Oxford : Pergamon, 1982, vol. 6, 2, 161-287.

Leclercq, D., Boxus, E., de Brogniez, P., Lambert, F., Wuidar H. (1993), The Taste approach: General implicit solutions in MCQs, open books exams and interactive testing. In: D. Leclercq, J. Bruno (eds.) Item banking, interactive testing and self-assessment. NATO ASI Series, Vol. 112. Berlin: Springer Verlag (this volume).

Leclercq, D. & de Brogniez Ph. (1990), A fresh look on confidence marking. In: Estes, Heene, Leclercq (eds.) New pathways to learning through educational technology. Proceedings of the Seventh International Conference on Technology and Education, Brussels, vol. 1, pp. 646-649.

Lichtenstein, S., Fischhoff, B., Phillips, L.D. (1975). Calibration of probabilities: the state of the art, decision making and change in human affairs. Proceedings of the Fifth Research Conference on Subjective Probability, Utility and Decision Making, Darmstadt, 1-4 September, D. Reidel.

Lindley, D.V. (1971). Making decisions. London: Wiley.

Luce, R.D., Raiffa, H. (1966). Games and decision. New York: Wiley.

Michael J.J. (1968). The reliability of a multiple choice examination under various test-making instructions. Journal of Educational Measurement 5, 307-314.

Miller, G.A. (1956). The magical number seven, plus or minus two. Psychological Review 63, 81-97.

Murphy, A.H., & Winkler, R.L. (1974). Subjective probability forecasting experiments in meterorology: some preliminary results. Bulletin of the American Meteorological Society 55, 1206-1216.

Pitz, G.F. (1974), Subjective probability distributions for imperfectly known quantities. In: Gregg, L.W. (ed.) Knowledge and Cognition. New York: Wiley, pp. 29-41.

Raiffa, H. (1970). Decision analysis, introductory lectures on choice under uncertainty. New York: Addison-Wesley.

Savage, L.J. (1951). The foundations of statistics. New York: Wiley.

Shannon, C.E. (1951). Prediction and entropy of printed English. Bell Syst. Techn. J. 30, 50-64.

Shuford, E., Albert, A. & Massengill, N.E. (1966), Admissible probability measurement procedures. Psychometrika 31, 125-145.

Shuford, E. (1993), In pursuit of the fallacy: resurrecting the penalty. In: D. Leclercq, J. Bruno (eds.) Item banking: interactive testing and self-assessment. NATO ASI Series F, Vol. 112. Berlin: Springer-Verlag (this volume).

Van Naerssen R.F. & Van Beaumont, R. (1965). Ervaringen met een Zekerheidsaanduiding bij objektieve Tentamens. Nederlands Tijdschrift Psychologie 20, 208-315.

Van Naerssen, R.F., Sandbergen, S. & Bruynis, E. (1966), Is de Utiliteitscurve van Examenscores een Ogief? Nederlands Tijdschrift Psychologie 21(6), 358-363.

The Development and Evaluation of ELI, an Interactive Elicitation Technique for Subjective Probability Distributions

Jelle van Lenthe

Department of Statistics and Measurement Theory, Faculty of Behavioral and Social Sciences, University of Groningen, Grote Kruisstraat 2/1, 9712 TS Groningen, The Netherlands

Abstract: Uncertain knowledge about continuous quantities is usually formalized through Subjective Probability Distributions (SPD's). Past research showed that the quality of SPD's is rather poor and that poor SPD quality might originate from method-induced biases. It is quite possible that equipped with more appropriate elicitation tools, assessors will prove to be more competent probability estimators than research thus far suggested. The present paper describes the search for a new elicitation methodology. The most important feature of the new ELIcitation technique ELI is the direct realization of a proper scoring rule in a graphically oriented interactive computer program. An evaluation study in which ELI was compared with other elicitation techniques showed that ELI performance was superior.

Keywords. Elicitation technique, uncertain knowledge, subjective probability distributions, proper scoring rules.

1 Introduction

To perform their analyses, decision and risk analysts usually need information about particular unknown quantities. Sometimes, insufficient data is available to determine the values interest objectively. In that case, often human subjects are called in as a source of information. For example, in risk assessment studies frequently the opinions of experts are used to assess the likelihood of rare, catastrophic events (Cooke, Mendel, & Thijs, 1988). Human knowledge generally is imprecise, uncertain, and more qualitative than quantitative in character. As input for a decision or risk analysis, however, the data are usually preferred in a quantitative mode. It is for this reason that the formalization of uncertain knowledge is an important topic in, for example, decision theory, risk analysis, and Bayesian statistics. Most studies deal with uncertainty related to discrete events (e.g., the likelihood of rain tomorrow). Less attention has been paid to subject of the present paper, the encoding of uncertainty related to continuous quantities (e.g., the percentage of rainy days next year).

Uncertain knowledge about continuous quantities is usually represented through probability distributions. The expression 'Subjective Probability Distributions' (SPD's) is used to emphasize the fact that these distributions reflect the subjective beliefs of an individual. The assessment of SPD's appears to be a rather difficult task for both statistically naive and statistically expert assessors. Several elicitation techniques supporting the specification of SPD's have been suggested. These techniques range from methods that directly ask for certain distribution characteristics to indirect procedures with a less clear relationship between response and final distribution. Detailed descriptions and discussions of pro's and con's of different techniques are given by, among others, Schütt (1981), Spetzler and Staël von Holstein (1975), and Van Steen and Oortman Gerlings (1988).

The research reported here was motivated by the desire to devise a completely new way of eliciting uncertain knowledge. The current paper, starts with the identification of quality criteria. Then these criteria are used to review the quality of the elicitation techniques. The conclusion will be that a new elicitation technique is required. The next section is devoted to the development of the new ELIcitation technique ELI. An operational description of ELI is followed by a discussion of the most important features of the technique. Then the main results of an evaluation study are reviewed. The conclusion is that ELI seems to be a promising alternative elicitation technique. Finally, future developments and research are discussed.

2 Quality criteria

Elicitation techniques should in the first place be judged on the quality of the SPD's they produce. From the point of view of assessors, SPD's are simply formal expressions of what they know or what they think to know. One might argue that it is not possible to evaluate SPD's in terms of right or wrong. However, Wallsten and Budescu (1983) showed that SPD assessment is much alike other forms of psychological measurement. Consequently, SPD's can be judged on the criteria of reliability and validity. In the second place, elicitation techniques can be judged on their practical usefulness.

The reliability of SPD's can be determined by assessing their stability in time or their consistency across methods. Based on the classical test theory model, Terlouw (1989) developed a triple model for the reliability of SPD's. With this model it is possible to consider the stability of items as well as the stability of persons. Correlations between repeated measurements are used to determine the stability of the central tendency, the width and the skewness of SPD's.

Strictly speaking, an SPD is valid when it accurately reflects the uncertain knowledge of an individual. Of course, it is very difficult to evaluate this internal validity and from a pragmatic point of view it is probably more important that SPD's correspond to events in the external world. This external validity of the assessments can be determined when the actual values of the uncertain quantities are, or eventually become, available. For SPD's to be externally valid they should have two properties: (a) SPD's as a group should be well-calibrated and (b) individual SPD's should be accurate.

SPD assessments are considered perfectly calibrated when, for all X, when the X%-credibility intervals of the SPD's have the property that X% of the actual values fall within those intervals. For example, for each of a large number of SPD's the 90%-credibility interval is calculated and then for each SPD it is verified whether the actual value falls within or outside the interval. Ideally, 90 percent of the actual values should fall within these intervals. The accuracy of individual SPD's can be evaluated in terms of a relationship between the SPD and the value that actually occurs. The accuracy measure should take account of (a) the difference between best guess and actual value and (b) the dispersion of the distribution. For example, a small difference between best guess and actual value should be accompanied by a small dispersion in order to achieve an optimal accuracy score.

An elicitation technique should not only contribute to the assessment of reliable and valid SPD's, it should also meet the requirement of practical usefulness. In the first place, an elicitation technique should be efficient; if two techniques produce equally reliable and valid SPD's, the more straightforward and fast method will be preferred. In the second place, an elicitation technique should be acceptable for different groups of assessors with various (statistical) backgrounds. So, different assessors should have no difficulty in learning and using the technique. In case of a computerized elicitation technique, the computer program should meet requirements of user-friendliness.

3 Quality of elicitation techniques

The above-mentioned criteria of reliability and validity of SPD's and practical usefulness of the technique are the touchstones for evaluating and comparing elicitation techniques. There has been relatively little research dealing with the problem of reliability. Whereas some attention is paid to the central tendency, hardly any study investigated the reliability of the dispersion of a distribution (Terlouw, 1989). The available data suggest that, in general, elicitation techniques are only moderately reliable (Lourens, 1984; Terlouw, 1989; Wallsten & Budescu, 1983).

More is known about the external validity and in particular the calibration of SPD's - for a review see Lichtenstein, Fischhoff, and Phillips (1982).

Numerous studies have shown that generally assessors are overconfident. Their SPD's tend to be much too tight and an unduly large percentage of actual values appears to fall into the extreme tails of the distributions. Often the percentage of actual values outside the 98%-credibility interval is used to evaluate overconfidence. This so-called surprise index should be 2 percent, but often values as high as 30 or 40 percent are observed.

Practical usefulness is another criterium which received only little attention - an exception is Schütt (1981). In general direct elicitation techniques appear to be more efficient, because the procedure is rather straightforward and inexpensive and does not require much time. Unfortunately, a straightforward procedure does not automatically constitute an assessment task that is easy for assessors. On the contrary, these direct techniques, which usually require values or probabilities as answers, appear to be difficult for statistically naive assessors. On the other hand, indirect techniques, which are in general less efficient, require hardly any statistical knowledge, for usually assessors are asked to choose between two or more alternatives.

From the recurrent observation of poor quality of SPD's, one might conclude that apparently existing elicitation techniques are inadequate. Thus far, however, only little attention is paid to the possibility that biases originate from the particular technique used (Fischer, 1982; Hogarth, 1980; Lourens, 1984). More frequently, cognitive or item-induced biases are considered responsible for the poor quality of SPD's. For example, Koriat, Lichtenstein, and Fischhoff (1980) showed that overconfidence can originate from the tendency of selectively focusing on evidence supporting a best guess and disregarding evidence contradicting it. Others demonstrated item-induced biases and showed, for example, that especially general knowledge questions frequently produce overconfidence (Wright & Wisudha, 1982; Ronis & Yates, 1987).

From previous research it is known that different techniques often elicit different distributions (Lichtenstein et al., 1982; Ludke, Stauss, & Gustafson, 1977; Van Steen & Oortman-Gerlings, 1988; Von Winterfeldt & Edwards, 1986). So, the technique itself appears to effect the assessments and poor quality of SPD's might be an artifact of the particular technique used. There is overwhelming evidence that particularly the elicitation technique most commonly used with continuous quantities - the fractile method - yields SPD's that tend to be much too tight (Alpert & Raiffa, 1982; Lichtenstein et al., 1982: Pickhardt & Wallace, 1977).

Reviews of evaluation and comparison studies show that, as far as the relative merits of elicitation techniques are concerned, the results have been contradictory (Lichtenstein et al., 1982; Seaver, Von Winterfeldt, & Edwards, 1978; Van Steen & Oortman Gerlings, 1988; Von Winterfeldt & Edwards, 1986). It is clear that no technique is uniformly best for assessing SPD's. Each elicitation technique has its own disadvantages and there apparently are no elicitation techniques that result in sufficiently reliable and valid SPD's.

In our opinion, poor quality of SPD's is at least in part an artifact of the particular techniques used. Human beings not only are step-wise information processing systems with limited capacities, they also are outstanding users of tools. Equipped with an appropriate tool they might be more capable estimators than research thus far suggested. In addition, as long as appropriate elicitation techniques are unavailable, it is difficult to draw unambiguous conclusions about cognitive- and item-induced biases. For these reasons, it was decided to develop an alternative and completely new way of eliciting uncertain knowledge.

4 Development of ELI

The quality criteria not only are appropriate touchstones for the evaluation of techniques, they also directed the development of the new ELIcitation technique ELI. Furthermore, several suggestions about improving the quality of SPD's and elicitation techniques found in literature (Hogarth, 1975; Hogarth, 1980, Huber, 1974, Lourens, 1984; Terlouw, 1989) guided the search for a new technique. These guidelines will be addressed after an operational description of the elicitation with ELI is given. In this way the most important guidelines can be illustrated at the new technique. Throughout the discussion the proportion will be assumed to be continuous quantity of interest.

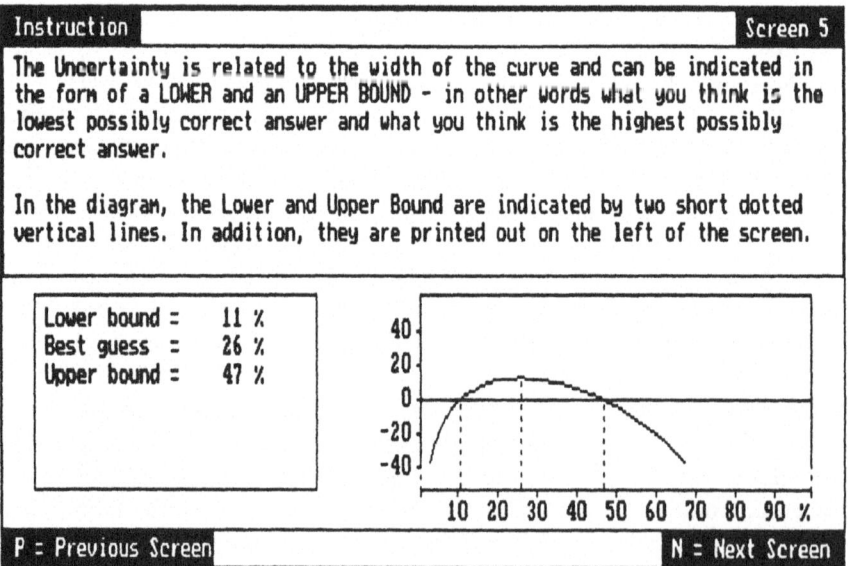

Fig. 1. The ELI curve represents a best guess and the uncertainty about it.

ELI is a graphically oriented interactive computer program and the ELI assessment task is a rather simple one. Assessors are not required to key in numbers. To assess their uncertain knowledge, all they have to do is manipulate a graphical curve with the cursor keys (see Figure 1). With this curve assessors can indicate their best guess as well as their uncertainty. The best guess corresponds with the top of the curve. The uncertainty is related to the width of the curve and is explained in terms of a lower and an upper bound. In other words, what assessors think is the lowest and what they think is the highest possible correct answer.

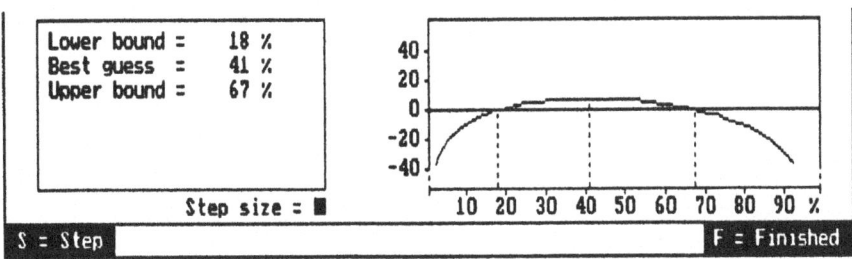

Fig. 2a. Uncertainty can be increased by using the down-arrow key.

The left-arrow and right-arrow keys control the horizontal position of the curve. The best guess can be indicated by shifting the curve to the left or the right. The up-arrow and down-arrow keys control the steepness of the curve. Uncertainty can be increased by using the down-arrow key. The curve grows flatter and the interval between lower and upper bound becomes larger (see Figure 2a). Using the up-arrow key reduces the uncertainty of the response. The curve grows more steep and the distance between lower and upper bound becomes smaller (see Figure 2b).

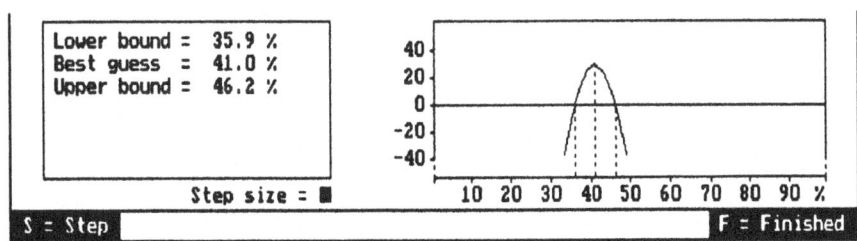

Fig. 2b. Uncertainty can be reduced by using the up-arrow key.

The curve not only represents a best guess and the uncertainty about this best guess, it also provides a score representation for uncertain knowledge. Assessors are informed that the curve provides feed-forward information about scores for all possible actual values. The score for a supposed actual value can be determined from the diagram by reading sideways from the curve to the vertical axis (See Figure 3). So, the curve reveals beforehand that positive scores will be obtained with actual values between lower and upper bound. Actual values outside this interval will yield negative scores. An ELI session typically consists of an interactive instruction, a few practice items, and the questions at stake. In the instruction part, assessors become familiar with the manipulation and score interpretation of the curve. If actual values of practice items are known, assessors can be provided with trial-by-trial outcome and score feedback (see Figure 3).

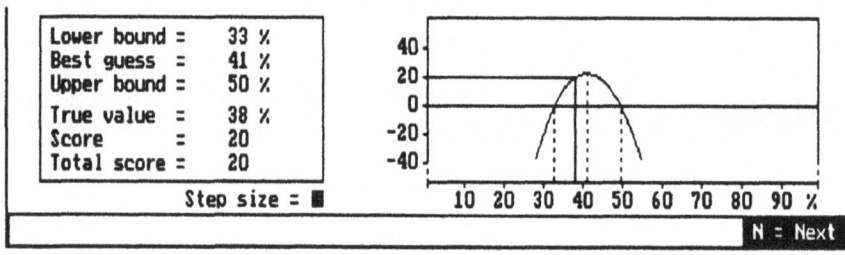

Fig. 3. When actual values are known, assessors can be provided with trial-by-trial outcome and accuracy feedback.

The manipulation of the graphical displays allows for a quick and convenient assessment. But, what exactly is assessed? This is best explained by introducing the main guideline - proper scoring rules should play a significant role in the new elicitation technique. Scoring rules provide the assessor with a score that reflects the correspondence between stated SPD and the value that actually occurs. During the assessment task (i.e. in the absence of knowledge of the actual value) the expected rather than the actual values are of primary interest. With a proper scoring rule assessors can maximize their expected score if and only if their SPD's correspond with their subjective judgments (Van Naerssen, 1961; Murphy & Winkler, 1970). Proper scoring rules are important elicitation tools for several reasons. First, announcing the use of proper scoring rules to evaluate the assessments can encourage assessors to report only true beliefs. In this way, proper scoring rules might stimulate the internal validity of SPD's. Second, proper scoring rules can be used afterwards for the actual evaluation of the external validity of the SPD. With proper scoring rules it is possible to assess the accuracy of a single SPD in terms of a relationship between stated SPD and the value that actually occurs (Murphy & Winkler, 1970; Winkler, 1986). Third, these accuracy scores might be useful in training in training situations for giving trial-by-trial accuracy feedback (Fischer, 1982; Staël Von Holstein, 1970, 1971).

It is obvious that ELI curves are not probability density functions. Yet, ELI assessments do result in subjective probability distributions. In ELI the family of beta distributions is used to represent uncertain knowledge about proportions. This family is chosen because (a) it is sufficiently rich (Terlouw, 1989) and (b) the beta distribution is the 'natural conjugate prior' for the proportion (Novick & Jackson, 1974). Assessors, however, do not deal directly with the probability distributions. They have to consider so-called score functions which, in fact, are transformed probability density functions. With this transformation, proper scoring rules come into the picture. Hofstee (1987) proposed to use the logarithmic scoring rule to transform the beta density functions $f_{p,q}(x)$. The resulting score function $ln(f_{p,q}(x))$ returns logarithmic scores for all possible actual values x. Figure 4 shows the density function $f_{p,q}(x) = x^{p-1}(1-x)^{q-1}/\beta(p,q)$ with $0 \leq x \leq 1$, $p > 0$, $q > 0$, and beta function $\beta(p,q)$ for $p=16$ and $q=11$ and the corresponding score function $ln(f_{p,q}(x))$. The zero score points of the score function correspond with the two points of intersection of the density function and the uniform distribution. These zero score points define quite naturally a lower and a upper bound; with actual values outside the interval between lower and upper bound, one loses to the ignorance assessment of the uniform distribution. For a more comprehensive description of a blueprint for the new elicitation methodology see Van Lenthe (1992).

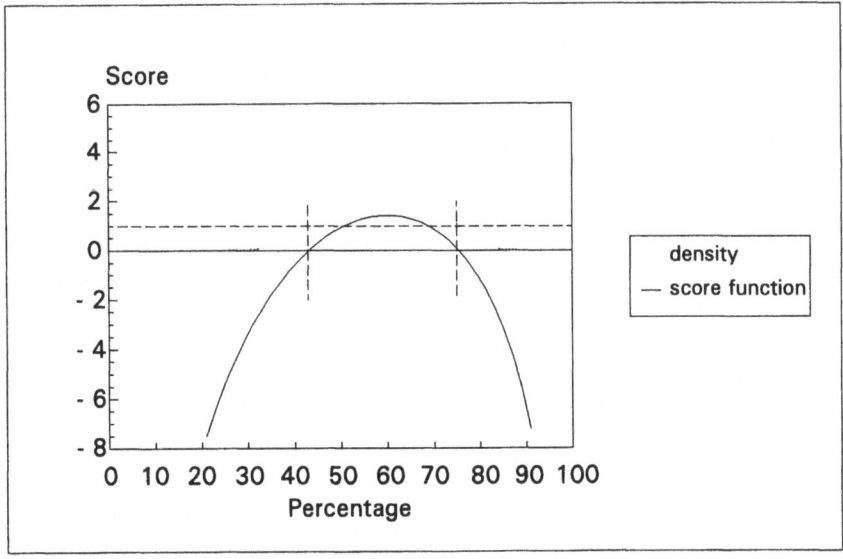

Fig. 4. The score function $S_{p,q}=\ln(f_{p,q}(x))$ and the underlying beta density function $f_{p,q}(x) = x^{p-1}(1-x)^{q-1}/\beta(p,q)$ with $0 \leq x \leq 1$, $p > 0$, $q > 0$, and beta function $\beta(p,q)$ for $p=16$ and $q=11$.

In ELI, the scoring rule is more than only an additional help. It is implemented as a central and regular part of the technique. Fortunately, other important guidelines were realized more or less automatically by the direct implementation of a logarithmic scoring rule in a graphically oriented interactive computer program. First, in its interaction with assessors elicitation techniques should preferably not use statistical concepts or methods and, considering the problems assessors have interpreting probability distributions, it might be better to use an alternative representation of uncertain knowledge. Studies of judgmental processes reveal that people have several shortcomings in acting as an 'intuitive statisticians' (Kahneman, Slovic, & Tversky, 1982; Hogarth, 1980). Consequently, people are ill-equipped to the task of assessing SPD's in the framework of the common statistical model. For this reason, the ELI assessment task only involves a best guess and the specification of uncertainty in terms of a lower and upper bound. Furthermore, in ELI a score representation of uncertain knowledge is used which is probably more compatible with the abilities of the human judge.

Second, an elicitation technique should place a minimum of information processing demands on the cognitive resources of assessors. Cognitive studies show that the human being is a selective, stepwise information processing system with limited capacity (Hogarth, 1975). So, an elicitation technique should allow the cognitive resources to be directed to the subject matter of estimating a continuous quantity. In other words, the elicitation technique should be straightforward, easy to handle and not difficult to learn. This consideration is realized by asking assessors to choose between alternatives rather than to key in numbers. The manipulation of the score curves allows for a quick and convenient assessment of the SPD's. Furthermore, we have tried to make the computer program is as user-friendly as possible. ELI is easy to handle and not difficult to learn and the program provides relevant information at appropriate stages of the elicitation process.

Third, an elicitation technique should encourage the reconsideration of assessments (Hogarth, 1975; Lourens, 1984; Terlouw, 1989). For that reason, an elicitation technique should stimulate assessors to reflect on the implications of their SPD's. In ELI reflection is stimulated by the score interpretation of the curves. The visualized feed-forward about possible consequences of an assessment might induce the assessor to reconsider the answer and perhaps in some cases to revise it.

Fourth, an elicitation technique should have a training option. On the one hand to become familiar with the technique. On the other hand training with trial-by-trial feedback might improve the quality of subsequent SPD assessments. When actual values of training items are known, ELI can provide trial-by-trial outcome and scoring rule feedback on the accuracy of the SPD. This scoring-rule feedback links up properly with the score interpretation of the curve. From the curve with possible scores, one is pointed out as the score actually obtained.

5 Evaluation of ELI

An experiment was carried out to evaluate ELI performance and to compare it with the performance of (a) the Subjective Probability Assessment Technique (SPAT) developed by Lourens (1984) and Terlouw (1989) and (b) the Simple Elicitation Technique (SET) which only asks for a best guess and a lower and upper bound. SPAT consists of a sequence of elicitation methods and seems to be one of the best available computerized elicitation techniques. SET was included because there is some empirical evidence that its three responses are as meaningful and informative as the assessments with more advanced methods (Moskowitz & Bullers, 1979; Terlouw, 1989). Furthermore, the simplicity of SET makes it probably an attractive alternative.

Forty-eight subjects participated in two sessions. In the first session they completed three item sets with 15 percentage questions each. The questions were based on quantities contained in a statistical almanac. Sequences of elicitation techniques and technique/item-set combinations were organized in a Greco-Latin square to control for item set effects and order and sequence effects of elicitation techniques. Estimations were repeated in a in a second session to provide for a measure of reliability of the techniques. After completing the item sets, subjects were asked to finish a questionnaire concerning the acceptability of the elicitation technique.

To evaluate and compare the elicitation techniques three criteria were employed: reliability of SPD's, external validity of SPD's, and practical usefulness of the technique. The triple model of Terlouw (1989) was used to explore the reliability of assessments. The external validity of the SPD's was examined with two calibration and one accuracy measure. Calibration was assessed by the percentage of actual values falling outside the 90%-credibility interval (our surprise index) and the percentage actual values falling within the 50%-credibility interval (the usual interquartile range). The logarithmic scoring rule was used to determine the accuracy of the SPD's. The practical usefulness of the techniques was evaluated by recording response times and by asking subjects to rate the task load and task support of the techniques.

The empirical results revealed that ELI support contributed to the assessment of reliable and externally valid SPD's. Moreover, ELI appeared to be an efficient and acceptable method. The quality of SPD's elicited with ELI turned out to be higher than the quality of SPD's elicited with either SPAT or SET. ELI and SET resulted in approximately equally reliable SPD's. But, compared with SPAT, the reliability of ELI SPD's appeared to be superior. In correspondence with past research results calibration scores with SPAT and SET pointed to a strong overconfidence bias. With ELI, however, overconfidence was almost absent and the surprise index and interquartile range scores were close to the optimal values of 10% and 50%. ELI support also caused the highest accuracy scores. The ignorance strategy of selection uniform distributions all the time would yield a zero score for each assessment. It is noteworthy, that only with ELI the mean accuracy scores were significantly higher than zero. The mean scores obtained with the two other techniques, in fact, were significantly worse than zero; an observation that corresponds with results of past research (Fischer, 1982; Lourens, 1984; Schaefer, 1976; Winkler, 1971; Yates, Zhu, Ronis, Wang, Shinotsuka, & Toda, 1989). Furthermore, compared with SPAT and SET, ELI turned out to be the most practical useful elicitation technique. ELI's cognitive support was rated much higher and its cognitive load was rated substantially lower. ELI also turned out to be a relatively fast method. It was only slightly slower than SET and much faster than SPAT. A more detailed presentation of the results of the ELI evaluation study is given in Van Lenthe (in press).

6 Conclusion and Discussion

The development of ELI was motivated by the notion that the recurrent observation of poor quality of SPD's might be attributed to the poor quality of existing elicitation techniques. It was anticipated, that equipped with a more appropriate elicitation tool assessors might prove to be more competent probability assessors than research thus far suggested. Several suggestions about improving SPD's and elicitation techniques guided the development of ELI. The cardinal guideline was that proper scoring rules should play a central role in the new elicitation method. They were used for the transformation of probability density functions in more manageable score functions. So, with ELI, assessors are not bothered about probabilities or densities, instead, they are asked to consider graphical displays of score functions. This direct realization of a proper scoring rule in a graphically oriented interactive computer program turned out to have some additional potential advantages.

Firstly, the score curve provides an alternative score representation of uncertain knowledge which probably is more compatible with the abilities of the human judge. Secondly, the visualized feed-forward information on possible consequences provided by the score curve, might stimulate assessors' reflection on the implications of their assessments. Thirdly, the curve provides a more natural operationalization for the often rather loosely defined concepts of lower and upper bounds. Finally, the scoring-rule feedback with training items links up properly with the score interpretation of the curve.

The objectives that guided the development of the new technique were realized to a great extent. ELI appeared to be an efficient and acceptable method and, more important, it contributed to the assessment of reliable and valid SPD's. These positive results encourage further development of the technique. The current ELI version was established mainly to study the merits of the central ideas for the new technique[1]. Consequently, thus far the range of possible applications limited. For instance, support is restricted to percentages (or proportions) and it is not possible to specify extreme degrees of certainty with the current version. Future advancement will be aimed at making the technique more flexible and more suitable for different applied and experimental settings. For example, the grid of possible curves will be extended allowing the estimation of very small percentages and the specification of more extreme certainties. Moreover, support will be not restricted to percentages, but also other quantities (e.g. means, correlations) and their corresponding natural conjugate probability distributions will be included.

In the current ELI version the course of an elicitation session is fixed and implemented by the designer of ELI. In the next version it will be possible for experimenters or decision analysts to implement their own instructions, their own training items, and their own questions of interest. In other words, ELI will support the construction of complete elicitation sessions. ELI is intended to support different groups of users. First, individual estimators who use ELI mainly as an elicitation aid, that is for the specification of their own uncertain knowledge. Second, experimenters or decision analysts who use ELI primarily as a tool to design an elicitation session. Third, subjects (e.g. substantive experts) who are asked to complete a particular ELI session.

Future empirical research will be aimed at exploring the most important reasons for the superior ELI performance. The technique is based on several suggestions from literature and some new ideas. So, there are several possible explanations; for example, (a) the new representation of uncertain knowledge with feed-forward about possible scores, (b) the graphical character of the technique, (c) the use of particularly the logarithmic scoring rule, etc.. At this moment, however, definite conclusions can not be given. In view of the positive experimental results thus far, it is expected that additional empirical research and further advancement of the technique will result in an elicitation tool that contributes to the assessment of reliable and valid SPD's. Moreover, it is anticipated that ELI will prove to be an efficient and for different users acceptable technique which, in addition, is suitable for different applied and experimental settings.

---

[1] The current ELI version runs on IBM-compatibles with a CGA, EGA, VGA, or Hercules graphics card. It appears to full advantage on an AT-machine (fast) and a VGA graphics device (resolution and colors). The further development of ELI will be a combined effort with the Interuniversity expertise center ProGAMMA. The final ELI-version and manual will be available begin 1993 from Iec ProGAMMA, P.O.Box 841, 9700 AV Groningen, The Netherlands.

Acknowledgements

This research was supported in part by PSYCHON-grant 560-266-014 of the Netherlands Organization of Scientific Research (NWO). The author thanks Ivo W. Molenaar and Willem K.B. Hofstee for their comments on earlier versions of the paper.

References

Alpert, M., & Raiffa, H. (1982) A progress report on the training of probability assessors. In D. Kahneman, P. Slovic, & A. Tversky (Eds.) *Judgment under uncertainty: Heuristics and biases* (pp. 294-305). Cambridge: University Press.

Cooke, R.M., Mendel, M., & Thijs, W. (1988) Calibration and information in expert resolution; a classical approach. *Automatica, 24* (1), 87-94.

Fischer, G.W. (1982) Scoring-rule feedback and the overconfidence syndrome in subjective probability forecasting. *Organizational Behavior and Human Performance, 29*, 352-369.

Hofstee, W.K.B. (1987, December) *Overconfidence*. (Available from Willem K.B. Hofstee, Department of Personality Psychology, Faculty of Behavioral and Social Sciences, Grote Kruisstraat 2/1, 9712 TS Groningen, The Netherlands.)

Hogarth, R.M. (1975) Cognitive processes and the assessment of subjective probability distributions. *Journal of the American Statistical Association, 70*, 271-289.

Hogarth, R.M. (1980) *Judgement and choice. The psychology of decision*. Chichester: Wiley.

Huber, G.P. (1974) Methods for quantifying subjective probabilities and multi-attribute utilities. *Decision Sciences, 5*, 430-458.

Kahneman, D., Slovic, P., & Tversky, A. (Eds.) (1982) *Judgment under uncertainty: Heuristics and biases*. Cambridge: University Press.

Koriat, A., Lichtenstein, S., & Fischhoff, B. (1980) Reasons for confidence. *Journal of Experimental Psychology: Human Learning and Memory, 6*(2), 107-118.

Lichtenstein, S., Fischhoff, B., & Phillips, L.D. (1982) Calibration of probabilities: The state of the art to 1980. In D. Kahneman, P. Slovic, & A. Tversky (Eds.) *Judgment under uncertainty: Heuristics and biases* (pp. 306-334). Cambridge: University Press.

Lourens, P.F. (1984) *The formalization of knowledge by specification of subjective probability distributions*. Dissertation, University of Groningen, Groningen.

Ludke, R.L., Stauss, F.F., & Gustafson, D.H. (1977) Comparison of five methods for estimating subjective probability distributions. *Organizational Behavior and Human Performance, 19*, 162-179.

Moskowitz, H. & Bullers, W.I. (1979) Modified PERT versus fractile assessments of subjective probability distributions. *Organizational Behavior and Human Performance, 24*, 167-194.

Murphy, A.H., & Winkler, R.L. (1970) Scoring rules in probability assessment and evaluation. *Acta Psychologica, 34*, 273-286.

Novick, M.R. & Jackson, P.H. (1974) *Statistical methods for educational and psychological research*. New York: McGraw-Hill.

Pickhardt, R.C. & Wallace, J.B. (1977) A study of the performance of subjective probability assessors. *Decision Sciences, 5*, 347-363.

Ronis, D.L. & Yates, J.F. (1987) Components of probability judgment accuracy: individual consistency and effects of subject matter and assessment method. *Organizational Behavior and Human Decision Processes, 40*, 193-218.

Schaefer, R.E. (1976) The evaluation of individual and aggregated subjective probability distributions. *Organizational Behavior and Human Performance, 17*, 199-210.

Schütt, K.-P. (1981) *Wahrscheinlichkeitsschätzungen im Computer-Dialog [Probability estimates with computer dialogue]*. Stuttgart: Poeschel Verlag.

Seaver, D.A., Von Winterfeldt, D., & Edwards, W. (1978) Eliciting subjective probability distributions on continuous variables. *Organizational Behavior and Human Performance, 21*, 379-391.

Spetzler, C.S., & Staël von Holstein, C.-A.S. (1975) Probability encoding in decision analysis. *Management Science, 22* (3), 340-358.

Staël von Holstein, C.-A. S. (1970) Measurement of subjective probability. *Acta Psychologica, 34*, 146-159.

Staël von Holstein, C.-A. S. (1971) Two techniques for assessment of subjective probability distributions. *Acta Psychologica, 35*, 478-494.

Terlouw, P. (1989) *Subjective probability distributions a psychometric approach*. Dissertation University of Groningen, Groningen.

Van Lenthe, J. (in press) ELI: an interactive elicitation technique for subjective probability distributions. *Organizational Behavior and Human Decision Processes*.

Van Lenthe, J. (1992) *A blueprint of ELI, a new method for eliciting subjective probability distributions*. Manuscript submitted for publication.

Van Naerssen, R.F. (1962) A scale for the measurement of subjective probability. *Acta Psychologica, 20*, 159-166.

Van Steen, J.F.J., & Oortman Gerlings, P.D. (1988) *Het gebruik van expertmeningen in veiligheidsstudies [The use of expert opinion in safety studies]*. Internal report, TU Delft/TNO.

Von Winterfeldt, D., & Edwards, W. (1986) *Decision analysis and behavioral research*. Cambridge: University Press.

Wallsten, T.S., & Budescu, D.V. (1983) Encoding subjective probabilities: A psychological and psychometric review. *Management Sciences, 29*(2), 51-173.

Winkler, R.L. (1971) Probabilistic prediction: some experimental results. *Journal of the American Statistical Society, 66*(336), 675-685.

Winkler, R.L. (1986) On "good probability appraisers". In P. Goel & A. Zellner (Eds.) *Bayesian inference and decision techniques* (pp. 265-278). Amsterdam: Elsevier.

Wright, G., & Wisudha, A. (1982) Distribution of probability assessments for almanac and future event questions. *Scandinavian Journal of Psychology, 23*, 219-224.

Yates, J.F., Zhu, Y., Ronis, D.L., Wang, D.-F., Shinotsuka, H., & Toda, M. (1989) Probability judgment accuracy: China, Japan, and the United States. *Organizational Behavior and Human Decision Processes, 43*, 145-171.

A Computer Environment to Develop Valid and Realistic Predictions and Self-Assesment of Knowledge with Personal Probabilities

Arie Dirkzwager

Huizerweg 62, 1402 AE Bussum, The Netherlands
E-mail: aried@ooc.uva.nl

Abstract : Knowledge is defined operationally by the ability to make predictions. In multiple choice tests one has to predict which alternative will be scored as the correct one. In this context a testee is called a predictor. A prediction is defined as the assignment of probabilities to mutually exclusive events. The "true" probabilities are dependent on the knowledgeability of the predictor, on the information he has: all probabilities are personal probabilities. Probabilities are called "true" if for each probability the proportion of occurrence of events approaches the probability assigned to those events. A measure of "realism" indicating if reported probabilities are an overestimation or an underestimation of the true probabilities will be derived as will be the scoring rule to evaluate predictors.

Organisms do adapt to their environment in such a way that the payoff from that environment to their behavior is maximized. So do human predictors. To shape the behavior of an organism one should shape it environment. Important characteristics of environments that grow trustworthy predictors (telling that they don't know for sure if they don't) can be derived mathematically and show interesting relationships with information theory. Paradoxically, the best measure of the predictor s' knowledgeability is not the optimal scoring rule.

It is quite feasible to build such optimal predictor environments with computers. The essence is that the utilities for the alternatives are variable: they have to be calibrated by the predictor. Examples using multiple choice tests and tests with open questions will be demonstrated.

Many-predictor systems are discussed as are the information-theoretic measures on the relevancy of questions asked in such systems. Differences with item analysis of educational tests are mentioned.

Keywords: Knowledge assessment, Personal probabilities, Confidence, Uncertainty, Information theory, Education, Decision making, Predicting, Computer based eliciting of knowledge, Computer based testing, Learning, Variable utilities, Open ended questions, Multiple choice, Policy.

1 Introduction

Knowledge can be defined operationally as the ability to predict something (de Groot 1960). A prediction is the assignment of probabilities to a set of mutually exclusive events of which one and only one will occur at scoring time. These events may be possible outcomes of some preplanned experiment but they need not. They may also be future observations of things yet "unknown" to the predictor. An example is the question of, which alternatives in a Multiple Choice test will be judged as "right" at scoring time. This should be "unknown" at testing time to those who take the test; at least their knowledge should be based on other information than information on the scoring key.

The statement that all probabilities are personal probabilities, that is to say that they depend on the information the predictor has on the events the probabilities are assigned to, needs not much arguing any more (see de Finetti 1970, Savage 1951). It is however quite central in what follows.

Predictions and expectations are important for rational decision making if the utilities of the different events can be estimated, but they determine also largely human behavior in general. Predictions can however also be elicited for the sole purpose of measuring the knowledge and knowledgeability of the predictor, in educational testing for instance. Shuford and Brown (1975) showed convincely that in an educational context this method, if properly implemented, is to be preferred and that Multiple Choice has some very undesirable side effects such as learning to study for shallow and superficial knowledge on many topics instead of studying for thorough knowledge on fewer topics (See also Dirkzwager 1987).

The predictor has some knowledge and expresses this knowledge in a probability distribution. From where this knowledge originates and what assumptions or theories the probabilities are based upon is not of prime interest, the validity of the predictions is, even if it is caused by *clair voyance* instead of solid scientific theory.

A prediction is called valid if, in a world where observations can be repeated the relative frequency of an event is in the long run equal to the probability assigned to that event. A predictor is called "realistic" if over all his predictions in the long run his assigned probabilities equal the relative frequencies of occurrance of events these probabilities are assigned to. A predictor who is completely ignorant and who reports consistently uniform probability distributions is very realistic, but most likely not very valid.

Predictions should be realistic and valid, how do we get such predictions? One method is to pool the predictions of "experts" in a computerized system according to some optimizing algorithm, but how do we spot the experts? Anyhow we can not do without individual human predictors for our input to the knowledge base of the expertsystem. An important educational question is how to raise valid and realistic human predictors. We may try to solve this problem by giving the appropriate instructions, but that's no solution as instructions are seldom payed sufficient attention to and when it is more profitable not to follow them they are neglected. The solution is to arrange the environment in such a way that valid and realistic prediction behavior is consistently rewarded and non-realistic and non-valid behavior is punished. With computers such an environment can and should be implemented. That humans in a test taking situation adapt to their environment and optimize their behavior in the direction of the optimal strategy given a certain scoring system is empirically shown by many authors, a.o. Leclercq (1982), Walsten and Budescu (1983), Lichtenstein and Fischhoff (1980), Dirkzwager (1981).

It might be clear that our philosophy is different from a psychometric approach of measuring personal feelings of belief, uncertainty, or confidence (like Walsten and Budescu 1983 and Terlouw 1989 discuss). That might be interesting to (clinical) psychologists but in our context it is crucial if uncertainty and confidence are warranted and realistic. Personal feelings may help or hinder predictors, we are only interested in the validity of their predictions (see also the discussion in 3.).

2 Measuring realism and the validity of predictions

We assume the difficult task of defining and formulating the context of the predictions done (see however 8.). We have a set of n well defined experiments or occasions j (j = 1,2,3,...n) to observe the occurrence of one and only one of a set of k mutually exclusive events i (i = 1,2,3,...k). The events need not be the same over all experiments, each experiment might be unique in this respect, only their number should be the same. The "experiments" might be a number of test items, the "events" being the alternatives for each item, they might be a specification of external conditions under which the observations could be done (time and place in the case of the weather forecast or the description of an experimental procedure leading to one of the predicted observations). The task of the subject is to make predictions, that is to say to assign probabilities to all events i on all occasions j. We call these reported probabilities responses r_{ij}.

2.1 Realism

Assuming that a subject made a large number of predictions we group the events according to the value of the assigned probabilities and note if the event is observed or not. Now in each group the proportion of observed events should be equal to the probability assigned to these events, it is an estimate of the true personal probability p the subject should have assigned to these events. If both are not equal the subject was not realistic in his reported personal probabilities. In that case he was either overestimating or underestimating his knowledge (or he was cheating). Following Shuford and Brown (1975) we model this as a linear relationship:

the reported probability r is a linear function of the true probability p:

(1) $$r_{ij} = \frac{1}{a} p_{ij} + b$$

From the fact that probabilities sum up to one follows:

(2) $$1 = \frac{1}{a} + kb$$

or:

(3) $$b = \frac{a-1}{ak}$$

where k is the number of alternatives.

Under this model the true probability equals:

(4) $$p_{ij} = ar_{ij} - \frac{a}{k} + \frac{1}{k}$$

This is an estimator of p_{ij} using the parameter a.

When we take N_r to be the number of events to which the probability r was assigned by the predictor and T_r the number of these events that was observed. Then we get a direct estimator of p:

(5) $$p(r) = \frac{T_r}{N_r}$$

To get a least squares estimator of the parameter a we should minimize:

(6 $$F(a) = \sum (p(r) - p_{ij})^2$$

where summation is over all responses r.

Substitution of (4) and (5) into (6) gives after some rearrangement:

(7) $$F(a) = \sum N_r (\frac{T_r}{N_r} - ar - \frac{1}{k} + \frac{a}{k})^2$$

where summation is over all observed values r of responses r_{ij}.

Rewriting (7) gives:

(8) $$F(a) = \sum N_r ((\frac{T_r}{N_r} - \frac{1}{k}) - (r - \frac{1}{k})a)^2$$

Expanding the square and differentiation gives:

(9)
$$\frac{\partial F(a)}{\partial a} = \Sigma \, N_r \, (\, 2(\frac{T_r}{N_r} - \frac{1}{k})(r - \frac{2}{k}) + 2(r - \frac{1}{k})^2 a)$$

We set this to zero and solve for a:

(10)
$$a = \frac{\Sigma(T_r r - \dfrac{T_r}{k} - \dfrac{N_r r}{k} + \dfrac{N_r}{k^2})}{\Sigma(N_r - \dfrac{2N_r r^2}{k} + \dfrac{N_r}{k^2})}$$

We observe that summation over all responses gives:

(11)
$$\Sigma N_r r^2 = \Sigma\Sigma r_{ij}^2, \; \Sigma N_r r = \Sigma\Sigma r_{ij} = n, \; \Sigma N_r = nk \text{ and } \Sigma T_r r = \Sigma\Sigma t_{ij}$$

where $t_{ij} = r_{ij}$ if event i is observed in experiment j and zero otherwise. Substitution in (10) gives:

(12)
$$a = \frac{k\Sigma\Sigma t_{ij} - n}{k\Sigma\Sigma r_{ij}^2 - n}$$

which is a measure for realism: if a = 1 all reported probabilities are "right", a<1 indicates overestimation of ones knowledge, a>1 underestimation (which seldom happens in reality). According to Shuford and Brown (1975) with a similar measure a quite stable estimate of a is reached already with 15 to 20 experiments (test items).

So far we assumed there is one constant value of a for each predictor independent of the topic he is predicting on and independent of his knowledgeability on the events he is predicting. It is quite improbable that this assumption is true: in cases the predictor is very knowledgeable he hardly can overestimate his knowledge, in case he "does not know" he only can overestimate, if he considers himself an expert in a certain field he probably tends to overestimate his knowledge in case he does not know precisely what to predict, whereas if it is not his field he does not care to admit he does not know and may even underestimate in his reported probabilities. People are calibrated differentially when dealing with items of varying degrees of difficulty. For experimental evidence see Lichtenstein and Fischhoff 1977 and Dirkzwager 1987a.

So the general parameter a is only a mean measure of more relevant parameters a_j dependent on the relationship of the predictor to the topic j he is predicting on. Predictions could be made more valid if we knew an estimate for these specific a_j's, but it is very doubtful if the validity would be raised by applying one general estimate a to all occasions j. We prefer to use the responded probabilities r_{ij} uncorrected for (lack of) realism.

If we still do correct, assuming the items leading to an estimation of the realism parameter are very homogeneous, this may result in negative probabilities and/or values larger then one in case of an underestimating predictor. Shuford and Brown propose to cut off and set the probability to zero resp. one in this case. A more elegant solution is to observe that for the item at hand it is impossible that the predictor underestimates his knowledge that much and set a at its maximum value that transforms the lowest reported probability for this item exactly to zero. By this method we are also sure that the probabilities sum up to one.

2.2 Validity

We will consider the most general case in which the universe behaves stochastically: we assume that on every occasion j an event i has the probability π_{ij} to occur. After the occasion we may observe that an event with probability π did occur and our personal probability p_{ij} is set to one, but even after the observation we may not be able to decide on one event. As far as I understand modern physics in some cases all that can be "observed" is a probability distribution over all events theoretically possible at the occasion and the most valid prediction is not the one that assigned a probability of one to the observed event (there is no such one event). The best one can do is to predict the probability distribution that will be observed.

A deterministic view of the universe is just a special case of the stochastic one in which all probabilities π_{ij} except one are set to zero. Then it is believed that for every occasion j there is one specific event i with probability $\pi_{ij} = 1$ that is bound to happen and all other thinkable events, however plausible before the observation is made, are in fact impossible ($\pi_{ij} = 0$ for those events). A multiple choice test with only one right answer is an example.

Whatever we believe about the universe in which predictions are made and validated, in any case a valid predictor reports probabilities r_{ij} that are equal to the "true" or "real" probabilities π_{ij}. Those real probabilities are unknown and when the occasion is unique in the sense that the experiment can not be repeated under exactly the same conditions they can not be estimated by computing relative frequencies. In most real life situations this is the case, take for instance the weather where we must abstract from a large number of variables to be able to say that the conditions are the same on two different occasions.

Still we will approach the problem of validating predictors by searching for a scoring procedure that gives the highest possible score to the most valid predictor, that is to say a predictor whose r's are all equal to the matching π's. We propose a weighted sum of scores, summation over events where the score for each event depends only on the probability assigned to that event. In formula:

$$(13) \qquad S = \sum \pi_i F(r_i)$$

We must chose F in such a way that S be maximum if all $r_i = \pi_i$ under the additional condition that $\Sigma r_i = 1$.

So we look for the maximum of:

(14) $S = \Sigma \pi_i F(r_i) - \lambda(\Sigma r_i - 1)$

Differentiation with the Lagrange multiplier λ gives k equations of the form:

(15) $\pi_i \dfrac{\partial F(r_i)}{\partial r_i} - \lambda = 0$

λ is some constant independent of r_i. So (15) can be rewritten as:

(16) $\dfrac{\partial F(r_i)}{\partial r_i} = \dfrac{\text{constant}}{\pi_i}$

For the best predictor $r_i = \pi_i$ for all i. In that case S should be at its maximum, so the derivative of $F(r_i)$ should be equal to some constant devided by r_i. This gives:

(17) $F(r_i) = A \ln(r_i) + B$

where A and B may be freely choosen constants.

Taking $-1/\ln(2)$ as the value of A and setting B equal to zero we get:

(18) $S = -\Sigma \pi_i \,^2\log r_i$

This scoring formula has an interesting information theoretic interpretation, it is the amount of information recieved by the predictor from the universe he observes. The universe is sending messages i each with a probability π_i. Each of these messages is expected by the predictor with probability r_i and has an information value equal to $-^2\log r_i$. So S is the mean information recieved. To the best possible predictor this is minimal and equal to the information sent: $-\Sigma \pi_i \log \pi_i$.

In a deterministic universe or after $\pi_i = 1$ is observed the score is only dependent on the probability assigned to the observed event, what the predictor said about events not observed is not relevant.

In the most common situations the universe may be stochastic but in most cases after the observation we can point to one and only one of the k events as the "true" one that occurred at this occasion. In that case the score for this item is equal to the logarithm of the probability assigned to this event. An estimate of the overall score S is the mean of these scores for a larger number of items:

(19) $S_{est.} = -\Sigma\Sigma \,^2\log t_{ij}$

it's expected value is given by (18).

If the predictor is realistic (see 2.1) we may assume π_{ij} equals r_{ij} and we may take his own mean expected score as a quicker and better estimate of his validity (Dirkzwager 1987b). It does not depend upon random events. Another advantage is that it can even be computed before scoring time:

(20) $$ES = -\Sigma\Sigma r_{ij} \,^2\!\log r_{ij}$$

it is equal to the amount of information he expects from the universe on base of his personal probabilities. It is equal to his personal uncertainty.

If "true" probabilities p_{ij} can be computed using the an estimate of the realism parameter a, a better estimate of the subjective expected score is:

(21) $$ES_{est.} = -\Sigma\Sigma \, p_{ij} \,^2\!\log r_{ij}$$

If we take this as a measure of the validity of the predictor we see that his validity, given the true probabilities p_{ij}, is maximum ($ES_{est.}$ minimum) if he is realistic and reports them truly with $r_{ij} = p_{ij}$ for all i,j.

2.3 "Knowing that one knows"

By learning to be a realistic and valid predictor (see par. 3), one learns to know what one knows and to know what one does not (yet) know. It is related to the ability to estimate if a given answer is correct or incorrect (the probability of being correct). In multiple choice situations this probability is at least equal to one divided by the number of alternatives, so for five alternatives at least 20%. With open questions this probability might approach to zero if one is forced to give an answer and does not know at all. Hosseini and Ferrell (1982) emphasize that it is important that students be able to estimate the correctness of their answers. They use the area under the ROC-curve from signal detection theory as a measure, quite another approach from that followed in this paper. There major objection is that our method is dependent on presenting a very small set of precoded alternatives with the guarantee that one of them is right. This however is not the case if we use open questions, which is quite well possible (see 5.2 and 6). The solution is to have a very large number of alternatives of which the contents are to be provided by the examinee, the last alternative being allways "None of these". The only restriction is that it should be enforced that non of the assigned probabilities be smaller than the probability assigned to this last alternative, otherwise one would get the maximum score by providing nonsense answers and assigning a probability of one to "none of these". A disadvantage of Hosseini and Ferrell's method is that no proper scoring rules are used and therefore their reward structure (see 3) does not reinforce true answers. A link from our approach to signal detection theory however would be interesting but the relationship is not yet clearly understood.

3 Reward structures

We want to design an environment in which realistic and valid predictors are rewarded more than less valid predictors and in which too unrealistic predictors are punished as their predictions are misleading and, dependent on the decision making context, possibly dangerous to rely upon. The reward should be the only determinant of the utility the effect of his predicting behavior has for the predictor, otherwise his behavior will (allso) be shaped by other motives that may interfere with becoming a valid and realistic predictor. Unwanted behavior should not pay in any way. It is a problem for psychologists to discover individual ideosyncratic motives and rewarding events and to explain and cure them if they are harmful. The most powerfull reward is probably if the predictor is payed (or has to pay) a substantial amount of money that is some linear function of his score.

We have to consider the possibility that opinions and decisions (political or other), that depend upon the predictions, have their own utility. Predicting is providing information. Some information can be very harmful, even if true, people might panic. Providing certain information might have very unwanted consequences. Making a good impression in front of less knowledgeable people might be a goal. In education the situation might be blurred if there is one cut off score and the most important criterium is pass or fail, or if the criterium is to be the best one and win a prize, situations discussed by Shuford and Brown (1975).

All those factors influence the predictions if they are not compensated for in any way. We will not go into this problem any further, but assume that society and education is organised in a such a way that the only motivation of the predictor is to maximize the score he gets, even if he is only able to get a quite low score. In that case maximizing his expected score is the best he can do. The problem then is how to compute this score.

We may derive the scoring formula using (13) to (17). $ES=(1/n)\Sigma S_j$ is the expected score if responses r_{ij} are given. This is maximum for $r_{ij} = \pi_{ij}$ if:

(22) $\qquad ES = \Sigma\Sigma\pi_{ij} (A \ln(r_{ij}) + B)$ $\qquad\qquad$ $(A>0)$

From this formula we take

(23) $\qquad S_{ij} = A \ln(r_{ij}) + B$

to be the score for item j if i is the right answer that is to say if $\pi_{ij} = 1$ (the occurring event on occasion j is event i).

Again setting $t_{ij} = r_{ij}$ if $\pi_{ij} = 1$ at scoring time (event i is observed) and zero otherwise the score should be computed as:

(24) $\qquad S = (1/n) \Sigma\Sigma (A \ln(t_{ij}) + B) (A>0)$

where A and B again are constants to be choosen freely.

We require that the maximum possible score for an item j be 100 points and the score be zero in case the prediction on item j was the uniform probability distribution (admitted complete ignorance). This gives:

(25) $A \ln(1) + B = 100$

and:

(26) $A \ln(1/k) + B = 0$ or $-A \ln(k) + B = 0$

Solving for A and B gives:

(27) $S = (1/n)\Sigma\Sigma(\dfrac{100}{\ln(k_j)} \ln(t_{ij}) + 100)$

and:

(28) $S_{ij} = 100 (\dfrac{\ln(r_{ij})}{\ln(k_j)} + 1)$

The question of which formula to use if we want to elicit _realistic_ answers $r_{ij}=p_{ij}$ is answered with the same reasoning replacing π with p in the foregoing formulas. It leads to the same scoring formula (27).

Mark that (27) is a special case of the more general scoring formula:

(29) $S = (1/n)\Sigma\Sigma\pi_{ij}(\dfrac{100}{\ln(k_j)} \ln(r_{ij}) + 100)$

which is applicable not only when only one event is observed, but also whenever the observation is a probability distribution π over events i. A problem here might be that the π_i's are also personal probabilities: equally able observators might still differ in their observations. This problem can be approached in the same way as the problem of how to pool predictions of different predictors we will discuss later on. The problem is how to combine the observations of different observators.

An objection often heard is that a predictor never can recover if he once assigned a probability zero ($r_{ij} = 0$) to an event that obviously might happen (π_{ij} not equal to zero, $\pi_{ij} > 0$) as his total negative score in that case becomes infinitely large. Hofstee (1980) is the only one I know of who does not mind, he calls this a case of a "festive bankruptcy": one never should say "never" or one's chickens will come home to roost some time. It is a clear case of overestimating ones own knowledge. Such overestimating however is in use in traditional empirical social sciences where a hypothesis is completely rejected if some event less probable than the "level of significance" happens. I propose to be lenient to this practice and take actual predictions with a pinch of salt by choosing for each predictor a realism parameter a somewhat smaller than one and by computing the score on basis of the corrected probabilities p_{ij}. If we take 1% as an acceptable level of significance $r_{ij}=0$ would mean $p_{ij} < .01$. From this the parameter a can be computed using (4):

(30) $\qquad .01 = 0 - \dfrac{a}{k} + \dfrac{1}{k}$

or:

(31) $\qquad a = 1 - .01\,k$

which leads by substitution into (29) to the scoring formula:

(32) $\qquad S = (1/n)\Sigma\pi_{ij}(\dfrac{100}{\ln(k_j)}\,\ln((1 - .01\,k_j)\,r_{ij} +.01\,) + 100\,)$

Where, dependent on the context and the seriousness of mistakes lower values then .01 can be taken for the significance level.

This method has as an advantage over a method that does not assign negative scores below a certain cut off point, that the properties of the probability distribution are maintained and that our leniency at the lower end goes with a certain mistrust at the higher end: very high probabilities are also "taken with a pinch of salt". A disadvantage is that some predictor who really knows is mistrusted too. The effect of this correction on the higher scores is however much less than on the large negative scores.

4 Reward structure and the validity of the predictor

The score according to (27) or (29) is not the best measure of a predictor's validity. Following the derivation of (20) and assuming the predictor is realistic a better estimate of his validity is his expected score:

(33) $\qquad ES = (1/n)\Sigma\Sigma\,r_{ij}(\dfrac{100}{\ln(k_j)}\,\ln(r_{ij}) + 100)$

It is not dependent on what event happens to realize at occasion j and it can be computed before any observation is made. As a matter of fact it is a linear function of the information theoretic uncertainty of the predictor. The more certain a realistic predictor is the more valid are his predictions. Paradoxically we cannot use (33) as the predictors score from which his rewards are computed. It's maximum is reached when the predictor pretends to be most certain, reporting a probability of one for any one event, no matter which one, and a probability of zero for the other ones.

5 A computer environment to elicit realistic and valid predictions

5.1 Characteristics of an optimal environment

An optimal environment to learn to predict realistically and validly should have the following characteristics:

1. The predictor should be clearly presented with the possible consequences of his predictions, that is to say when he is adjusting his probabilities for each event i it should be computed for him what his score-would be if that event i occurs. This idea of feed forward was launched for the first time in 1975 (Dirkzwager 1975). It provides a good method to get well calibrated predictions and to prevent "overconfidence" (van Lenthe 1992).

2. The predictor's task of assigning probabilities should be facilitated as much as possible:
2.1 Administrative computing to make the sum of his reported probabilities equal to one should be done for him, when he changes one probability the others should automatically be proportionally adapted.
2.2 If it is not necessary to report a specific probability for all events and if it suffices to report only the probability of the (subjectively) most probable event, that should be all that is asked for. That is possible if the most probable event is the one that occurs. Only if it didn't the predictor is informed that he was "wrong", the probability of this event is fixed and can not be changed any more, and he is asked to give the probability of the next most probable event until the probability of the occurred event is found. If the predictor thinks it easier to estimate first the lower probabilities (excluding very improbable events by assigning a very low probability to them) he should be able to do so.

3. Feedback should be as immediate as possible.
3.1 When the predictor has decided on the probabilities he assigns to the events, and as soon as it is known which event is the true one, he should be informed if it is the event he thought most probable and if so what his score on this item is. If not, he may adapt the probabilities of the remaining events within the remaining bounds (see 2.2). This procedure is repeated until the item is scored.
3.2 The realism measure "a" (12) should be computed and updated after each occasion on which an item is scored. After the score is reported the predictor should be informed immediately if he is sufficiently realistic or if he is either overestimating his knowledge and should be more careful in assigning his probabilities or if he is underestimating and should dare to report more extreme probabilities. If the predictor is "moving in the right direction" (the parameter "a"

going to one) this feedback could be misleading and should not be given.

A computer program called "TESTME" with those characteristics was developed and is available for the Macintosh, MS-DOS machines and the Atari ST1040. It has also the facility to input "tests", sequences of items j with alternatives i. As it is now, the test author has to indicate in advance which alternatives are the "correct" ones to make immediate and automatic scoring and feedback possible.

5.2 Open questions

The method can be used for open questions too. If we follow the sequential procedure described in Sect. 3.1 we may leave the contents of the alternatives open. The subject has to formulate the possible answers himself, starting with the most probable one. He assigns a probability and is allowed to formulate the next best possible answer (alternative 2) with its probability, assuming or being informed that the first one was wrong. The probabilities must be in descending order. The number of guesses (alternatives) allowed should be fixed beforehand and quite large (say $k = 50$ with a probability of only 2% for any answer if the subject really does not know). The last alternative always is "None of these": the right answer was not found in the foregoing $k-1$ tries. If the subject cannot think of any (other) alternatives (any more) he reports uniform probabilities (for the remaining ones). All answers given should have a subjective probability larger than the probability of "none of these", otherwise it would pay to provide nonsense answers and assign a probability of one to "none of these".

This can be quite easily implemented in an interactive computer program eliciting answers and probabilities, providing we have some procedure to match answers given with the right answer automatically or by human judgment.

5.3 Immediate feed back on predictions

With real predictions some on-line (network) system would be needed that gives immediate feedback to all predictors as soon as it is known which event of the prediction occurred. Such a system could also be used to compute one best prediction from the individual predictions taking into account the scores and the estimated realism of the predictors. This best prediction could be used as a (provisional) criterion to score the individual predictions even before the observations are mad. An item bank that is continuously updated is a special case of such a network: with many "predictors" or subjects relevant item characteristics can be estimated (sum of scores and their variance for each alternative assuming it were the right one, see Sect. 8.3.).

6 Variable utility tests instead of multiple choice

In fact in the environment described in Sect. 5 the assignment of probabilities r_{ij} is the real task, it is not only an intermediate step in assigning personal utilities S_{ij} to the different events. These utilities may vary over predictors and occasions or items. As a matter of fact we may leave out the probabilities completely, presenting only the possibility to vary the S_{ij}:

$$(34) \qquad S_{ij} = 100(\frac{\ln(r_{ij})}{\ln(k_j)}+1)$$

keeping the conditions $0 < r_{ij} < 1$ and $\Sigma r_{ij} = 1$ automatically. The only restriction to the predictor or testee is that Sij cannot be higher than 100. What should be done by the computer environment whenever a S_{ij} is changed by the predictor is solving (34) for r_{ij}, adapting all r_{kj} and computing and presenting all new S_{kj} (k # i). Predicting and taking a test becomes a gambling game with a positive expected reward if one bets reasonably according to one's informedness. In this game the subjects (predictors or testees) can determine themselves their stakes between zero points whatever event happens (no betting) to 100 points *if* a choosen event happens with (k - 1) times the risk of losing $(100(\ln(0.1)/\ln(k)+1)$ points (see (32)). For two alternatives (k = 2) this may amount to risking -520 points, no reasonable bet if one is not very very sure of the right alternative.

The personal probabilities need not be reported but can be derived afterwards by solving (34). Then they can be used to determine, e.g., the realism of the predictor. This is a situation quite different from traditional multiple choice where the stake is allways the same and the best thing one can do is to pick the alternative with the highest personal probability however small this may be. Testing with variable utilities provides more precise information. An interactive computer program that implements this paradigma with open questions (see Sect. 5.2) will be demonstrated.

7 Many predictors

If we have many predictors all specifying their personal probabilities realistically over one and the same set of events of which only one will be observed at scoring time, we have the problem how to get one optimal prediction out of them. Taking for each event some mean probability would not do, as in that case the opinion of the least knowledgeable predictor would weigh as much as the opinion of the experts and the resulting uncertainty would be larger than the uncertainty of the most valid predictor.

Taking the probability distribution of the most valid predictor according to the measure of (un)certainty (20) seems a good alternative but has two objections against it. First any one predictor might try to dominate the final result by pretending zero uncertainty. Second two equally certain and realistic predictors may still differ substantially; which one should be choosen in this case? Obviously we need some procedure to assign weights to the individual predictions, but I don't know any such procedure that is derived following some sound mathematical argumentation.

We may adopt the following reasoning if we want to pick one alternative as the true one. In a testing situation this would amount to deciding upon a scoring key that maximizes the sum of all scores and thus is the most profitable scoring key to the testees.

The method is simple and it is based on the observation that, with realistic predictors, the most knowledgeable predictor would (and could) risk a large negative score S_{ij} (28) for highly improbable events i to make a good chance of getting a substantial positive score. Assigning a certain probability r_{ij} to an event is to agree upon getting the score S_{ij} if that event happens to be the case. Less knowledgeable predictors choose less extreme scores S_{ij}, in the case they know nothing they should assign the same score zero to all alternatives, they cannot discriminate between them. If they are cheating or overestimating their knowledge high and low bets S_{ij} can not be distributed over the events in a valid way. So summation of the possible scores S_{ij} which are determined by the predictors themselves gives for each event i a total payoff P_i of which the variance is systematically determined by the most valid (realistic and certain) predictors:

(35) $\qquad P_{ij} = \Sigma S_{ij}$, summation over predictors.

Less knowledgeable predictors do not substantially contribute to this variance or their contribution is non-systematic error variance in case they are cheating or overestimating. The best guess on the most probable event is on the one with the highest total payoff and this event could be taken to be the "true" one in the scoring key.

It is not yet clear how these total payoff's should be transformed into a proper probability distribution. Rigorous proof is lacking that the resulting r_{ij}'s would be optimal and that they are maximizing (22), which is the criterion for an optimal predictor. Anyhow the total pay off is the pay off the group of predictors as a whole sets at stake. We could compute corresponding probabilities by solving (36) for r_i.

(36) $\qquad P_i = 100 \left(\dfrac{\ln(wr_i)}{\ln(k)} + 1 \right)$

where w should be choosen such that $\Sigma r_i = 1$. This gives:

(37) $\qquad wr_i = \exp\left(\left(\dfrac{P_i}{100} - 1 \right) \ln(k) \right)$

with:

$$(38) \qquad w= 1/\Sigma exp((\frac{P_i}{100} - 1) \ln(k))$$

I was not able to proof or disproof that this method results in a realistic and valid composed predictor, it may or may not depending on the context of the predictions, especially on the question if the different individual predictors are independent.

In general if one assigns a probability of say 16% to a certain event (quite probable) and one is informed that many different experts independently also judge the event "quite probable", say 66%, one is justified to raise one's probability. That is the use of asking a second opinion in medical diagnosis for instance. On the other hand if all predictors base their predictions on the same information and the same theoretical considerations, for instance in the case of predicting whether a dice will show 5 or 6, probability about 66%, that remains the "right" prediction even if it is confirmed by many "independent" predictors. If one believes in extrasensory perception or clairvoyance however one may even in this case adapt one's predictions if some other predictors assign higher or lower probabilities. So I think there is not one best algorithm to compute a "best" realistic and valid composed prediction. Different proposed algorithms for different specific contexts should be validated experimentally in the same way any predictor is validated by using proper scoring rules.

A non mathematical method would be to make the predictors discuss the prediction and negotiate on one common prediction that should determine their reward. In that case the "predictor" would be such an organized group of individuals, each with his specific knowledge and information they can share. In that case one may hope that group members learn to judge what information and arguments are new and should change the different probabilities and what information is all ready taken into account and should not be of influence any more. In classroom situations this set up entails often very edifying discussions that contribute to the individual learning processes. It would be an interesting experiment to elicit the individual predictions beforehand and compare the negotiated stakes with the total pay-off for each event computed from the individual predictions. My conjecture is that in groups that are heterogeneous with respect to information background they correlate highly whereas in homogeneous groups probabilities and negotiated stakes will be inflated. According to Seaver (1978, cited in von Winterfeld and Edwards 1986, p.133) group probability assessments are no better than some average of individual assessments because of this inflation when redundancy among group members is overlooked. When this inflation is found it should be empirically checked whether it is justified, by application of the proper scoring rule (24).

8 Questions and answers: the context of predictions

8.1 Applicability dependent on the kind of questions asked

The methods presented have their restrictions. They are clearly only applicable when the questions are formulated in such a way that it is possible to decide at some moment which answer is the right one. With open ended questions this means they should ask "who ....?", "when ....?", and "what ....?". Much more difficult to judge are answers to questions after the "why?" or "how?" if no multiple choice answers are precoded. If the question is asking for a short essay about some topic (a very good method to elicit knowledge) it cannot be judged "right" or "wrong" but it should be graded, and proper scoring rules could only be applied to a student's prediction of his grade. This however would only measure how far he knows if he knows, not the level of his knowledge of the topic. Still, for the judgement and the assignment of grades, a system using proper scoring rules could be used. In this case judging examiners should answer with probabilities to a multiple choice question like "This essay was written by: A. An expert in the field, B. A very good student. C. A good student. D. A student with poor knowledge. and E. Someone who knows nothing of the topic at hand."

Another restriction seems to be that answers should be on a nominal scale, questions like "What is the mean age at which Dutch women have their first child" are to be answered with a probability distribution over a continuous numerical scale. One might say that some answers near the true mean are "more right" than others with a small deviation that cannot be judged completely wrong. Staël von Holstein (1970) following Murphy (1969) proposed a quite complicated proper scoring formula for this situation and proved it sensitive to the distance of an answer to the "most" right answer. For these kind of questions however van Lenthe (1992) presents a more elegant solution using the logarithmic proper scoring rule. He also designed a quite efficient computer program to elicit these probability distributions interactively.

It is worthwhile investigating if his method is also applicable when the answer requested is a probability distribution over probabilities. As it is now with disjunct events on a nominal scale, only one value for their probabilities are to be reported. It would be more realistic to ask for a probability distribution. It would have a very small or no variance for events like coin tossing (the probability of getting heads *is* 50% and no other value) but a substantial variance for other events having a probability of about 50% (say the probability that Bush will be reelected). In the case of mutually exclusive events (multiple choice questions) it would be interesting to study the theoretical relationship between the different simultaneous probability distributions for these events. That would give a connection between van Lenthe's method and the one presented here.

8.2 Asking relevant questions

So far we assumed the question as given. Knowledgeability however shows more in asking the proper questions than in answering the questions others prepared. So in a really open situation the predictor should formulate not only the answers, but also the questions. Then the problem arrises of discriminating good questions from irrelevant ones. In a network of predictors this concerns the input of questions on which predictions are based and observations are to be made. Our theory may contribute to solving this problem.

A good and relevant question should have two characteristics in relation to the predictor:

1. The uncertainty on what the right answer would be should be large at the start of the process of observation and scoring:

$$(39) \qquad U = -\Sigma\Sigma \, p_i \,{}^2\!\log(p_i) \text{ (summation over predictors and alternatives)}$$

should be large, see (18).

2. The expected information one would get in the process of observation and scoring should be large, which amounts to the same criterion if we substitute the best guess p_i for π_i in (18). According to this criterion the best question would be one with an uniform probability distribution over a very large number of alternatives: a sheer gambling situation. This clearly is not what is wanted. Instead of criterion 2 we should require that the information any other predictor than oneself gets from answers to the question should be maximized. The possible answers and their most probable probabilities (according to the predictor who invented the question) are fixed. He may ask for the predictions q_i of another predictor. The information that predictor would receive according to the one who asked the question is:

$$(40) \qquad -\Sigma \, p_i \,{}^2\!\log(q_i)$$

and should be maximized.

This criterion amounts to requesting that the discrepancy between any two predictors should be large, the maximum being reached if they assign a probability of 100% to different alternatives.

With many predictors, say M, we have a M x M matrix for pairs of predictors telling for each predictor what information he may expect if the other one's probabilities are the same as in the universe from which the observations are taken. Summed over rows and divided by M this gives the mean expected information according to one specific predictor p. Summed over columns and divided by M this gives the mean expected information to be recieved by a specific predictor q. The sum total divided by M^2 seems to be a good measure for the relevance and the importance, for the quality Q of the question:

$$(41) \qquad Q_j = -\Sigma_i \; \frac{\Sigma p_i}{M} \; \frac{\Sigma \, {}^2 log \, (q_i)}{M}$$

This should be low for relevant questions for which observations can be expected to provide much information to a large number of predictors. If for these observations costly experiments are needed it might be wise to use this criterion for the decision which experiments should have priority: that would be crucial experiments in fields where different competing theories are available but no decision on the best one can be made (we assume predictors have some "theory" to justify their predictions).

With relevant questions predictors widely diverge in their probabilities and the mean probability for any alternative approaches $1/k$. Substitution in (41) gives:

$$(42) \qquad Q_j = -(1/k)\frac{\Sigma\Sigma \, {}^2 log \, (q_{ij})}{M} \qquad \text{(summation over predictors and alternatives)}$$

which is a linear function of the total sum of possible scores, (34) and (35).

From (42) follows quite naturally to take ΣQ_j (summation over all questions a person proposed for inquiry) as a measure of someone's knowledgeability and (scientific) creativity.

In (educational) knowledge testing this measure is less applicable than when case questions are asked to which the answers are allready known to the experts: quite uninteresting questions from the present point of view.

8.3 Questions for (educational) testing

In (educational) knowledge testing we ask questions to which the right answers are known. The purpose is to discriminate between subjects according to their level of knowledge in a certain field. To investigate if we succeeded in asking the right questions to that purpose an item analysis is performed. With the proposed proper scoring rule and scoring method we can get more information than with traditional item analysis based upon the multiple choice or true-false items. For each alternative (instead of each item) we get a score for each subject on a numerical scale. Principal Components Factoranalysis would show what is the factor measured and how well the different alternatives load on this factor. With good questions there is one (right) alternative with a large positive loading and the remaining alternatives have large negative loadings. We may take the mean and the variance of the right alternative as measures of item difficulty and discriminating power and the variance of the mean scores on the remaining alternatives as a measure of how well the distractors are choosen (this variance should be low). Sound (statistical) theory tailored to this testing method and item analysis however is largely lacking and should be developed.

9 Applications, implementation and political actions

The litterature on proper scoring rules seems to have died out. Around the year 1980 we find quite a few scientific publications on the topic and the strong arguments to apply them to educational testing are well known, be it mainly among the scientifically trained supporters, but not among the educators and policy makers in education. They are quite happy with Multiple Choice Tests, they rather forget the arguments raised against them when they were first introduced on a large scale, and they resist discussing a paradigm that is new to them. Supporters and researchers of methods based on proper scoring rules seem to be frustrated and leave the topic. That is regrettable. Many interesting questions remain for study, research and development as I hope I made clear.

But even while many questions remain to be answered it is clear that the methods based on proper scoring rules are to be preferred in (educational) knowledge testing and even may have larger utility in other contexts of prediction and decision making. Hofstee (1980) proposed to make it the foundation for the "empirical discussion" in (experimental) social sciences and Bernardo (1979) also proposed to choose for experiments that maximize the expected information to be gained from it. We made it plausible that the method proposed here may help funding agencies to choose the most relevant research for support. Military intelligence could profit from a multiple prediction system to assess information, economic enterprises could also be assisted in their planning by these prediction systems and policy could be better underpinned if it used them. If these organisations (and society as a whole) were organised in such a way that the scores of the predictors determined their pay-offs, we would have a realistic and valid cognitive foundation for decision making where natural selection would give the most knowledgeable people the strongest voice. This is quite feasible with the present computer networking and communication technology.

A problem is that policy makers should be convinced. They are apparently not by articles in scientific journals or by whatever comes to the fore at conferences among specialists. The general public should be informed and political lobbying activity is needed. Large scale implementation in education would be a good start, but than large institutions like Educational Testing Service in the USA and the CITO in the Netherlands should be convinced first.

In the meantime we should be alert for unwanted negative side effects. From the experience with Multiple Choice and from our own theory on prediction we know that they have a probability larger than zero, even if we do not yet see them. If anyone can formulate them and predicts them with a substantial personal probability, it would be a very relevant experiment to investigate if he is right or wrong (see 8.2), for I think a majority of experts in the field assigns a probability near zero to them.

References

Bernardo, J.M. Expected Information as Expected Utility. Annals of Statistics 1979,7,686-690.

de Groot, A.D. Methodologie. The Hague. 1960, Mouton.

de Groot, A.D. Research as a Learning Process and Policy as a Learning Environment. SVO Workshop on Educational Research and Public Policy making, 1981, The Hague.

De Finetti, B. Logical Foundations and the measurement of Subjective Probabilities. Acta Psych. 1970, 34, 129-145.

Dirkzwager, A. Computer-based Testing with automatic scoring based on Subjective Probabilities. Computers in Education, eds. Lecarne, O. Lewis, R.; IFIP North-Holland Publ. Company 1975, 305-311.

Dirkzwager, A. Een democratische en objectieve beoordeling van de wetenschappelijke betekenis van een experiment. Congres van Psychologen 1981 Nijmegen, Intern rapport vakgroep Functieleer en Methodenleer, Amsterdam 1981, Vrije Universiteit.

Dirkzwager, A. Education for Knowledge Evaluation and Information Assessment. 2ᵈInt. Conf. on "Children in the Information Age", 1987ᵃ, Sofia.

Dirkzwager, A. Personal Probability Testing and the Assessment of Knowledge, a new pardigma: Multiple Evaluation. Europ. Meeting of the Psychometric Society, 1987ᵇ, Enschede.

Hofstee, W.K.B. De Empirische Discussie, Meppel 1980, Boom.

Hosseini, J. Ferrell, W.R. Detectability of Correctness: a Measure of Knowing that one Knows. Instr. Science 1982,11, 113-127.

Leclercq, D. Confidence Marking: its use in Testing. Evaluation in Education, 1982, 6, 161-287.

Lichtenstein, S. Fischhoff, B. Do those who know more also know more about how much they know. Org. Beh. and Human Performance 1977, 20, 159-183.

Lichtenstein, S. Fischhoff, B. Training for Calibration. Org. Beh. and Human Performance 1980, 26, 149-171.

Savage, L.J. The Foundations of Statistics. New York 1951, Wiley.

Shuford, E. Brown, T.A. Elicitation of Personal Probabilities and their Assessment. Instr. Science 1975, 4, 137-188.

Staël von Holstein, C-A.S. Measurement of Subjective Probability. Acta Psych. 1970, 34, 146-159.

Terlouw, P. Subjective Probability Distributions: a Psychometric Approach. Groningen Univ. 1989, dissertation.

van Lenthe, J. ELI: An Interactive Elicitation Technique for Subjective Probability Distributions. In: Org. Beh. and Human Decision Processes, december 1992.

van Lenthe, J. Scoring-rule Feedforward and the Elicitation of Subjective Probability Distributions. Groningen University, 1992: submitted for publication.

van Lenthe, J. The Development and Evaluation of·ELI, an Interactive Elicitation Technique for Subjective Probability Distributions. In: D. Leclercq, J. Bruno (eds.): Item Banking: Interactive Testing and Self-Assessment. NATO ASI Series F, Vol. 112. Berlin 1993, Springer-Verlag (this volume).

von Winterfeldt, D. Edwards, W. Decision Analysis and Behavioral Research. Cambridge 1986, Cambridge University Press.

Wallsten, T.S. Budescu, D.V. Encoding Subjective Probabilities: a psychological and psychometric review. Management Science 1983, 29, 151-173.

Weltner, K. The Measurement of Verbal Information in Psychology and Education. Berlin 1973, Springer-Verlag.

The Dependability of Test Scores: Generalizability Theory and Hierarchical Linear Models

Alexander Renkl

Institut für Empirische Pädagogik und Pädagogische Psychologie, Universität München, Leopoldstr. 13, D-80802 München, Germany

Keywords: Reliability, generalizability theory, hierarchical linear models, measurement of change.

Introduction

Generalizability theory (Cronbach, Gleser, Nanda & Rajaratnam, 1972; Shavelson, Webb, & Rowley, 1989) is an extension of classical reliability theory. It provides a useful and flexible framework for dealing with the dependability of psychological and educational measurement, especially with regard to observational or cognitive process data. Also with respect to test items, generalizability theory provides a useful framework for addressing some important questions concerning the quality of measurement. However, generalizability theory is not suited for dealing with systematic changes over time or over instructional units (e.g., learning gains). For the latter purpose, hierarchical linear models are very powerful statistical tools. Thus, in the first part of this paper some uses of generalizability theory are presented; in the second part the application of hierarchical linear models in modelling change is discussed.

1 Generalizability theory

In contrast to classical or probabilistic test theory, in the framework of generalizability the influence of *several* error sources can be simultaneously analyzed (e.g., inhomogeneity of items and temporal instability). These error sources are called *facets*. The dependability (reliability) is considered as the generalizability of scores over the *universe* of possible conditions of the facets. For example, a score is reliable if it can be generalized over different occasions and over different items. The first is know as retest reliability in classical test theory, the latter is know as internal consistency. Whereas in the classical approach, these aspects must be separately treated, one advantage of generalizability theory is that the problems of temporal stability and of item consistency can be simultaneously addressed.

1.1 ANOVA model of generalizability theory

Generalizability theory can be described within an ANOVA framework. A fictitious study in which the influence of the kind of items and of the test occasions on a test score is analyzed serves as an illustrating example. In the case of two error sources, that is, two facets, a three-factorial model results; the third factor represents the persons. If the realized conditions of a facet can be regarded as random sample out of a theoretically infinite universe, the facet is considered as random factor (for fixed factors, see below). Table 1 shows the ANOVA model.

Table 1.
Decomposition of observed scores for a two-facet, crossed design

---

	$X_{pio} = \mu$	$\sigma^2 X_{pio} =$
Main effects		
person	$+ \mu_p - \mu$	σ^2_p
item	$+ \mu_i - \mu$	$+ \sigma^2_i$
occasion	$+ \mu_o - \mu$	$+ \sigma^2_o$
Interaction effects		
person x item	$+ \mu_{pi} - \mu_p - \mu_i + \mu$	$+ \sigma^2_{pi}$
person x occasion	$+ \mu_{po} - \mu_p - \mu_o + \mu$	$+ \sigma^2_{po}$
item x occasion	$+ \mu_{io} - \mu_i - \mu_o + \mu$	$+ \sigma^2_{io}$
three-way interaction	$+ (X_{pio} + \mu_p + \mu_i + \mu_o$	$+ \sigma^2_{pio,e}$
+ residual variance	$- \mu_{pi} - \mu_{po} - \mu_{io} - \mu)$	

Note p: person; i: item; o: occasion; e: residuum.

Except for the grand mean μ that is, the overall mean, each component has a corresponding distribution. For example, for persons (p), there is a distribution of $(\mu_p - \mu)$ with a mean of zero und a variance of $E (\mu_p - \mu)^2 = \sigma^2_p$. This variance is called *universe score* variance and corresponds to true variance in classical reliability theory. The distribution of the remaining (error) components have a mean of zero, as well, and - demonstrated by the item facet - a variance of $E (\mu_i - \mu)^2 = \sigma^2_i$. Table 1 shows the components that contribute to the variance of single scores $(\sigma^2 X_{pio})$:

(1) Person effect: To what extent does a person deviate from the grand mean, that is, what is the achievement level of a person?

(2) Item effect: To what extent does an item deviate from the grand mean, that is, how difficult is it?

(3)   Occasion effect: Is the level of scores on an occasion lower or higher than usual?

(4)   Interaction between persons and occasions: Are the persons differently rank ordered[1] on different occasions?

(5)   Interaction between persons and items: Are the persons differently rank ordered on different items?

(6)   Interaction between items and occasions: Are there different rank orders with regard to item difficulty on different occasions?

(7)   Three-way interaction between persons, items, and occasions: To what extent is the score dependent on the specific combination of person, item, and occasion?

The residual variance and the three-way interaction are confounded, because persons are regarded as factor and, therefore, each cell has only one observation (n = 1).

To sum up, the variance of the (non-aggregated) scores is made up of the seven non-confounded variance components that can be distinguished in a two-facets crossed design (Table 1). The magnitude of the various components informs about, how much each error source contributes to lacking dependability of measurements. Numerical estimates for these variance components can be obtained by computing the mean squares for each component and then by solving a set of simultaneous equations. These equations can be found in Cronbach et al. (1972).

1.2 Coefficient of generalizability

The coefficient for the dependability of scores, called generalizability coefficient, is defined analogous to the reliability coefficient in classical test theory, that is, as percentage of universe score variance (σ^2_p) comprised in total variance ($\sigma^2_p + \sigma^2_{error}$):

$$E(\rho^2) = \frac{\sigma^2_p}{\sigma^2_p + \sigma^2_{error}} \qquad (1)$$

The error variance is additively made up of the variance components due to the error facets. However, the specific definition of the error term depends on several aspects:

(1)   The kind of test norm: criterion- vs norm-referenced test, or in terms of generalizability theory, absolute vs relative decisions

(2)   The amount of aggregation, for example, over items or over occasions

(3)   The intended scope of generalization.

In the following, these points will be elaborated.

---

[1]Technically, an interaction effect can also occur when the rank order remains constant and just the distances change. However, practically significant interaction effects include a change in rank order.

1.3 Relative versus absolute decisions

Generalizability theory distinguishes two cases, namely absolute and relative decisions. If one is merely interested in interindividual differences, as in testing under a norm-referenced perspective, relative decisions are made. In absolute decisions, one is concerned with the absolute level of scores, as in criterion-referenced testing. Measurement error σ^2_{error} is differently defined for each of these purposes.

In the case of *absolute* decisions, such as in criterion-referenced testing, all variance components, except for the universe-score variance, contribute to the error term:

$$\sigma^2_{error\text{-}abs} = \sigma^2_i + \sigma^2_o + \sigma^2_{pi} + \sigma^2_{po} + \sigma^2_{io} + \sigma^2_{pio,e} \qquad (2)$$

In the case of *relative* decisions, a smaller (or at maximum an equal) error term results, because the main effects of the facets are not included. An example may illustrate this point: it makes a difference whether one uses easy or difficult items if one is interest in whether a student masters a specific content area or not, as in criterion-referenced testing. It does not make any difference if one is merely interested in interindividual differences, as long as the variations in item difficulty merely result in a higher or lower achievement level for all persons (main effect) and no person x item interaction emerges. Thus, in the error term for relative decisions, only the interaction effects are included:

$$\sigma^2_{error\text{-}rel} = \sigma^2_{po} + \sigma^2_{pi} + \sigma^2_{io} + \sigma^2_{pio,e} \qquad (3)$$

The main point is that different error terms and, therefore, generalizability coefficients result depending on the type of intended decision, that is, whether one is interested in interindividual differences or in the absolute level as in criterion-referenced testing.

1.4 Necessary amount of aggregation: economical design planning

Up to now, we have considered the generalizability of a single score, that is, of a person's response to one item at one occasion. In order to increase the dependability of measurement, a score is often constructed by aggregating over several conditions of a facet, for example, over several items. This means reduces the error due to the item facet. The effect of test prolongation on reliability is well-know. In mathematical terms, the error variance components have to be divided by the number of conditions (n_i) over which it is aggregated. For example, if it is aggregated over seven items $(n_i = 7)$, the variance components due to the item facet (main and interaction effects) are divided by seven. This procedure parallels the rationale of the Spearman-Brown formula in classical reliability theory. Thus, the full definitions of the error terms for absolute and relative decisions are as follows:

$$\sigma^2_{error\text{-}abs} = \sigma^2_i/n_i + \sigma^2_o/n_o + \sigma^2_{pi}/n_i + \sigma^2_{po}/n_o +$$
$$+ \sigma^2_{io}/(n_i n_o) + \sigma^2_{pio,e}/(n_i n_o) \qquad (4)$$

$$\sigma^2_{error\text{-}rel} = \sigma^2_{pi}/n_i + \sigma^2_{po}/n_o + \sigma^2_{io}/(n_i n_o) + \sigma^2_{pio,e}/(n_i n_o) \qquad (5)$$

The main effects are divided by the number of conditions over which it is aggregated and the interaction effects are divided by the product of the number of conditions. Thus, the amount of reduction in error variance that is achieved by varying amounts of aggregation can be determined. This possibility cannot only be used to compute the generalizability of scores in the current design, but also to predict the dependability of scores, if the number of items or occasions is changed. In addition, one can determine the effects of increasing or decreasing the number of items and of simultaneously increasing or decreasing the number of occasions. Thus, generalizability theory is a useful tool for efficient design planning. It can help to determine the minimal amount of data collection that is necessary for reliable measurement.

1.5 Scope of generalization

Another advantage of generalizability theory is that different coefficients are obtained depending on the intended scope of generalization. If one wants to generalize over items and occasions, seven components enter into the error term, at least when interest in absolute decisions. However, if one does not want to generalize over different occasions, that is, if one is merely interest in the score at a certain moment, the error components that are associated with occasions can be omitted in the error term. An analogous modification of the error term results if one is merely interested in generalization over occasions, but not over items.

The last three sections showed that the definition of the error term depends on the purpose of measurement, on the amount of aggregation, and on the intended scope of generalization. So far we have discussed the generalizability over facets that can be regarded as random factors. However, there are also facets with a fixed set of possible conditions. The next section will deal with this topic.

1.6 Random versus fixed facets

Whereas items and occasions can be regarded as random sample out of a theoretically infinite population or, at least, as interchangeable (Shavelson et al., 1989), this is not true for any possible facet. For example, in interactive testing, there is only a limited number of possible or common types of item formats (e.g., graphical and verbal). The presentation format may influence the achievement measurement with regard to the absolute achievement level (main effect of format). Furthermore, there may be a person x format interaction, because different formats impede or favor student with different cognitive or learning styles.

If we focus on a limited set of conditions, the corresponding factor (facet) is considered as "fixed". Generalizability theory treats fixed facets such as item format differently than random facets. Rather than asking how many conditions are needed for reliable measurement, generalizability theory examines whether it is reasonable to average over the conditions of a fixed facet, that is in our example, over different item formats or whether separate scores should be computed for each item format. In order to answer this question, Shavelson et al. (1989) propose to temporarily regard a fixed facet as random and then to determine the error variance associated with this facet. If a large variance component is associated with the main and interaction effects that are related to this facet, that is, to the item format, separate scores should be computed for each item format. If a small proportion of variance is associated with item format, it is reasonable to aggregate over different formats.

To sum up this point, generalizability theory can help to find the proper universe of generalization, that is, whether one should aggregate over the fixed conditions of a facet and accordingly interpret the resulting score or whether a more differentiated view must be adopted, that is, separated scores have to be computed.

1.7 Flexibility of defining the object of measurement

Cardinet, Tourneur, and Allal (1981) proposed to extended generalizability theory with respect to research questions that do not regard persons as objects of measurement. For example, in interactive testing, it may be important to determine the difficulty of items. If one is interested in the reliability of difficulty estimates, the variance due to the main effect of items is regarded as universe score (or true) variance, whereas variance associated with persons must be considered as error. Cardinet et al. (1981) also present some computational procedures to handle the case of a fixed facet, such as item format, as differentiation dimension, that is, as source of universe score variance. Thus, for example, the reliability of differences in difficulty of various item formats can be determined.

Furthermore, Cardinet et al. (1981) provide a solution to another problem of traditional generalizability analyses. If different item formats put an advantage at students with a certain cognitive or learning style, this item x cognitive style interaction variance is attributed to a person x item interaction and, therefore, as error variance in traditional analyses. Cardinet et al. (1989) demonstrate possibilities - at least if it seems reasonable to define discrete groups of subjects with a certain preferred learning style - to create scores where this variation does not contribute to the error term.

To sum up this section, extensions of generalizability theory have been developed that allow a great flexibility in defining the dimension of differentiation. Merely some applications of this possibility were demonstrated in this paper; for an intensive discussion of this topic see Cardinet et al. (1981).

The uses of generalizability theory presented in this paper are the following:
(1)   Different reliability coefficients depending on the kind of decision (relative vs absolute)
(2)   Prediction of reliability coefficients for different designs (economical design planning)
(3)   Different reliability coefficients depending on the intended generalization (e.g. over time, over items)

(4)   Provision of guidelines for defining the appropriate universe of
        generalization
(5)   Flexibility in defining the object of measurement

1.8 A serious limitation of generalizability theory

Generalizability theory has a serious limitation if it is used for analyzing the
dependability of the measurement of learning gains. For example, if a score is
generalized over occasions, generalizability theory makes the assumption that the
measurement objects remain constant over occasions. This is called the *steady
state* assumption of generalizability theory. Thus, generalizability theory does not
distinguish between systematic trends and unsystematic variance over occasions.
In addition, it does not discriminate between differential growth curves for different
persons and unsystematic person x occasion interaction. A powerful statistical
approach for dealing with developmental trends or learning gains is the use of
hierarchical linear models.

2 Hierarchical linear models

The use of hierarchical linear models for measuring change is a very promising
approach. Due to space limitations, merely a brief overview over the use of
hierarchical linear models in the measurement of change can be given.

2.1 The rationale

Originally, the development of hierarchical linear models was instigated by the
problem that the nesting of students within, for example, classrooms caused
serious difficulties for the common statistical procedures, because the subjects
within the same classrooms usually have more in common than subjects from
different classrooms. Therefore, the number of independent observations does not
equal the number of subjects as it is normally assumed in doing inferential
statistics. In recent years, statistical models have been developed for this problem.
Bryke and Raudenbush (1987) showed that the rationale of hierarchical linear
models cannot only be applied if persons are nested within classrooms, but also if
there are multiple measurements within persons. In this case the measurements
within a person correspond to the students within a classroom and the persons
correspond to classrooms.

2.2 The steps

For the measurement of change, the agenda of using hierarchical linear models is
as follows: At the first, or within-subject stage, it is determined what kind of
growth trajectory adequately describes the average change of the sample, for
example, whether there is a linear or an accelerated, that is, a quadratic growth
curve in achievement. At the next stage it is tested whether the interindividual
differences in initial status and in change (growth trajectories) are significant. Of
course, one can also test the significance of interindividual differences in final
status. In addition, reliability estimates can be obtained for interindividual

differences with regard to differential status and with respect to differential growth trajectories. Thus, in learning environments, the significance of interindividual differences in pre-knowledge and in learning gains can be determined. Furthermore, in contrast to usual correlational techniques where the correlation of initial status or learning gains is biased by the regression effect, hierarchical linear models provide an estimate of the "true" correlation between initial status and learning gains. Thus, it can be determined to what extent students with higher or lower pre-knowledge profit form the learning environment. Finally, interindividual differences in initial status and in change can be related to personal characteristics (e.g., intelligence, cognitive style, test anxiety, achievement motivation) or to experimental treatments.

The use of hierarchical linear models in measuring change can be summed up by the following points:

(1)   Modelling the kind of growth trajectory (linear, quadratic/accelerated etc.)
(2)   Significance and reliability of interindividual differences with regard to initial status
(3)   Significance and reliability of interindividual differences with regard to change (growth trajectory)
(4)   Estimation of the true correlation between initial status and change.
(5)   Relating personal characteristics or experimental treatments with interindividual differences in initial status or change.

2.3 The regression models

In order to get a better idea how hierarchical linear models work, a brief look at the underlying regression models is useful. At the first stage, where it is determined what kind of growth trajectory adequately describes the average change of the sample, we have the following within-subject model (for the sake of clarity, the following regression models are slightly simplified):

$$y = b_0 + b_1 * x + b_2 * x^2 + error \qquad (6)$$

y: achievement or knowledge state
x: time in the instructional program or number of instructional units (0..t)
b_0: initial status
b_1: parameter representing the degree of linear growth
b_2: parameter representing the degree of accelerated or delayed, that is, quadratic growth.

In contrast to usual regression models, the parameters b_0, b_1, b_2 are not necessarily regarded as valid for the whole sample or population. It is rather tested whether the persons significantly differ with respect to these parameters. Interindividual differences in b_0, b_1, b_2 correspond to interindividual differences in initial status, linear growth and quadratic growth, respectively. On the between-subject level, potential interindividual differences in the parameters b_0, b_1, b_2 are regressed on student characteristics or experimental treatments:

$$b_0 = \beta_0 + \beta_1 * z_1 + \beta_2 * z_2 + ... + \beta_n * z_n + error \qquad (7)$$

$$b_1 = \beta_0 + \beta_1 * z_1 + \beta_2 * z_2 + \ldots + \beta_n * z_n + \text{error} \qquad (8)$$
$$b_2 = \beta_0 + \beta_1 * z_1 + \beta_2 * z_2 + \ldots + \beta_n * z_n + \text{error} \qquad (9)$$

$\beta_0 \ldots \beta_n$: regression weights for the between-subjects model

$z_0 \ldots z_n$: student characteristics or experimental treatment

In this way, causes or, at least, correlates of differential learning gains can be determined. A limitation of hierarchical linear models is that at least three measurement points are necessary. The data collected in simple pretest post-test designs cannot be analyzed in the described way. On the other hand, hierarchical linear models can cope with varying numbers of measurements and varying time points of measurement within the persons. In that respect they are extremely flexible.

3 Conclusions

The usefulness of both generalizability theory and hierarchical linear models was demonstrated in this paper. These approaches do not exclude or contradict but supplement each other. Generalizability analyses can be performed to determine the reliability or generalizability of scores, if the steady state assumption is reasonable, that is, if one wants to generalize over conditions of facets without considering systematic change. However, if change, for example, learning gains should be assessed, hierarchical linear models provide the possibility to estimate the reliability of the measurement of change. Furthermore, interindividual differences in change can be related to causes or correlates. The difference between generalizability theory and hierarchical linear models with regard to reliability estimation is that, metaphorically spoken, generalizability theory looks into the scores at the within-subject level and decomposes the aggregation over, for example, items in order to determine the generalizability of the aggregated scores. In contrast, hierarchical linear models provide reliability estimates at a higher level, that is, at the between-subject level. Thus, the dependability of interindividual differences in growth parameter can be estimated.

Unfortunately, despite their advantages, both approaches are relatively seldom used. Modelling change by hierarchical linear models is a recent evolution in applied statistics. Thus, it may need time to get know within the research community. In contrast, the basics of generalizability theory were developed almost 30 year ago (Cronbach, Rajaratnam, & Gleser, 1963). Maybe, the often very technical presentations have put off researchers to use this approach. I hope that my presentation has created some interest in these statistical models and that, in the future, they will be more often used in analyzing test data.

References

Bryke, A. S., & Raudenbush, S. W. (1987). Application of hierarchical linear models to assessing change. *Psychological Bulletin, 101*, 147-158.

Cardinet, J., Tourneur, Y., & Allal, L. (1981). Extension of generalizability theory and its applications in educational measurement. *Journal of Educational Measurement, 18,* 183-204.

Cronbach, L. J., Gleser, G. C., Nanda, H., & Rajaratnam, N. (1972). *The dependability of behavioral measurements: The theory of generalizability for scores and profiles.* New York: Wiley.

Cronbach, L. J., Rajaratnam, N., & Gleser, G. C. (1963). Theory of generalizability: A liberalization of reliability theory. *British Journal of Statistical Psychology, 16,* 137-163.

Shavelson, R. J., Webb, N. M., & Rowley, G. L. (1989). Generalizability theory. *American Psychologist, 44,* 922-932.

Human Self-Assessment:
Theory and Application to Learning and Testing

Darwin P. Hunt

Human Performance Enhancement, Inc., Executive Center II, 345 North Water Street, Las Cruces, New Mexico 88001, U.S.A.

"The proud man...is an extreme in respect of the greatness of his claims, but a mean in respect of the rightness of them; for he claims what is in accordance with his merits, while others go to excess or fall short."

"...he who thinks himself worthy of great things, being unworthy of them, is vain."
Aristotle, 4th Century B.C.

Introduction

In 1962 I taught my first university course - introductory statistics. I prepared my lectures well and delivered substantive and interesting lectures with the purpose in mind of the students learning the material from these lectures, the textbook, a workbook, etc. To my almost-terminating disappointment, the performance of the students on the first test was extremely poor. Their poor test performance came as a surprise to me because, based upon their questions, class attendance, apparent alertness during class periods, etc., I believed that they were learning the material adequately.

   Largely due to my immediately prior 11 years experience as an ergonomics/ engineering psychologist working on human-machine aerospace systems, I conceptualized my classroom teaching role as shown in Figure 1.

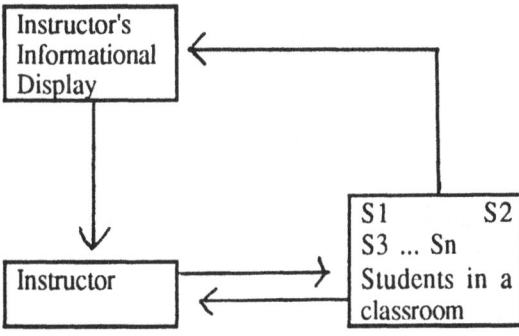

Figure 1. The classroom teaching situation as a closed loop system.

An analysis of this learning and teaching problem led me to the conclusion that the main problem was that my interpretation of my observations of the students' behavior, e.g., their facial expressions, etc. was inaccurate. Namely, often I believed that they were understanding the lecture material when, in fact, they were not. Thus, an improvement in the accuracy (validity) of the feedback information to me, the lecturer, should improve the effectiveness of the instruction.

To improve the feedback data, the classroom was instrumented so that each student was given a hand-held button, which he/she was told to press when they were not understanding the lecture. These buttons were connected by wires to a meter mounted on my lecture podium which displayed the percentage of students who were pressing their buttons at any moment in time. When the needle on the meter reached too high of a percentage, say 20%, then I would stop and try to remedy the apparent lack of understanding.

It should be noted that the students must make an assessment of whether they understand or do not understand the lecture in order to decide whether to press or not to press the button. A consideration of the ability of people to make these kinds of self assessments led me into an extensive program of research and thought concerning (a) the human processes which might underlie this activity and (b) the practical implications of these self assessments for learning and performance.

One of my first concerns was that requiring a student to perform this secondary task of self assessment might interfere with the primary task of classroom learning. For example, the additional information processing and other associated responses demanded by the performance of the self assessment task while a person is engaged in acquiring new responses may interfere with learning.

To test this hypothesis I conducted a preliminary laboratory experiment involving the task of learning a sequence of digits in which (a) a control group simply learned the sequence by repetition and (b) an experimental group was required to indicate, each time an answer was given, whether they were sure that their answer was correct. To my surprise, the experimental group which was required to perform the secondary task of self assessment learned the material in significantly fewer trials compared to the control group. Since this was a preliminary experiment and hastily conducted, it was not submitted for publication. At about this time I accepted the department headship at New Mexico State University which diverted my energies to administrative tasks and prevented my continuing this line of inquiry.

Approximately ten years later I was able to return to my interest in self assessment and conducted a well controlled experiment to examine the effects of self assessment (SA) responding on paired-associates learning (Hunt, 1982). The number of trials required to learn the names of the eight hand tools by people who made SA responses using either two, four, or eight SA-response buttons, was compared with the number of trials required by subjects in control groups who made no self assessments. Also, half of the SA subjects executed their answer response first followed by their SA response; the other half of the subjects executed their SA response first followed by their answer. There was a total of 180 subjects in the various experimental and control groups.

The results showed that learning was expedited by as much as 25% by SA responding, i.e., subjects using 8 SA response categories required significantly fewer trials (15.3) than did a control group which simply gave answers (20.5

trials). Also, learning was more rapid if the SA response was executed after rather than before the answer.

The finding that learning is affected by the order in which the answer and the SA response are executed may give some insight into how to interpret the observation that SA responding enhanced learning. To pursue this line of thought, a signal detection analysis was employed.

In the signal detection analysis of these learning data, one can conceptualize the subjects self assessment task as that of deciding whether a weak "signal of knowing" is present or absent. It is assumed that the SA responses of "Sure" and unsure provide reasonable estimates of the person's decision that the "signal of knowing" is present or absent. The accuracy of the SA responses may be estimated by the calculation of the hit rate (HR) and false alarm rate (FAR) as the conditional probabilities of p(Sure|Correct) and p(Sure|Wrong), respectively.

In short, this analysis found that subjects were better able to discriminate between knowing and not knowing the correct answer if they gave the answer first, followed by the SA response (dr = 0.85 vs 0.52). Furthermore, this greater sensitivity is due entirely to a higher HR (0.61 vs 0.49) rather than the FAR (.30 vs .30). The higher HR suggests that if a correct response has been covertly selected, then its execution helps the learner to confirm its correctness. The finding that the FAR is not affected by the order of response execution suggests that the execution of a wrong answer has no affect on the accuracy with which it is identified by a person as the wrong answer.

I report this experiment in some detail because it was the first experiment which I conducted on what I now call "human self assessment process theory" and this initial thinking set the direction of most of our subsequent thinking and applied work.

Table 1. The mean Hit Rate (HR), False Alarm Rate (FAR), and Sensitivity (dr) for the two orders in which the answer (R) and self assessment (SA) responses are executed.

execution	Hit Rate	False Alarm Rate	dr Sensitivity
R-SA	0.61	0.30	0.85
SA-R	0.49	0.30	0.52

Misinformation and usable knowledge

The idea of false alarms leads to the idea of misinformation, i.e., being sure of something which, in fact, is wrong. The practical importance of misinformation is that people make decisions, and select and execute actions based upon knowledge about which they are sure. Thus, from an educational and training point of view it is important to detect and identify such misinformation. Once identified, instructional approaches can be implemented to try to remedy the erroneous knowledge.

In the usual test of knowledge or achievement, it is determined whether a person can answer questions correctly. If a person gives the wrong answer, no distinction is made between whether the person was sure or unsure of its correctness.

However, in many situations, the distinction between being uninformed and misinformed is important. For example, if we are administering a licensing test to a professional (say a physician), it is important to make the distinction between whether the potential professional (a) has little confidence in the wrong answer which he gave and is not likely to use the knowledge in his practice or (b) has a strong belief that the answer is correct and is highly likely to use the erroneous knowledge in making (medical) decisions.

Similarly, in the usual test of knowledge, if a person gives a correct answer, it is assumed that the person has that specific knowledge (or perhaps was correct by guessing). However, no distinction is made between whether the person is sure or unsure of his correct answers.

It is commonly accepted that people behave in accordance with their knowledge; the more confident the belief then the more reliable is the response. If a person is sure of the

Criterion: Stockholm is the capital of Sweden		
Person's knowledge is that Stockholm is the capital		Person's knowledge is that Karlstad is the capital
Correct		Incorrect
Sure	Unsure	Sure
Informed (to some level)	Uninformed	Misinformed
Usable Knowledge	Unusable Knowledge	Usable Knowledge

Figure 2. The relationship among the correctness of a person's knowledge, the sureness with which the person believes the knowledge to be correct, and the usability of the knowledge (adapted from Hunt & Furustig, 1989).

correctness of erroneous knowledge, then the performance of tasks which rely on this misinformation will likewise be in error.

This line of thought leads to the concept of "usable" knowledge, in contrast to unusable knowledge. Usable knowledge means that a person is sufficiently sure of the knowledge so that it will be used to make decisions, and to select and execute actions. Unusable knowledge is knowledge about which a person is not sure enough to use it as a basis for deciding or acting.

Figure 2 illustrates some relationships among correctness, sureness, and usability of knowledge. Usable knowledge may be either correct or incorrect. In either case the person is considered to be usably informed because the person would use the knowledge, correct or incorrect, as the basis for acting. If the knowledge is correct, then the person may be considered well informed or, simply, informed. If the knowledge is incorrect, then the person is considered to be misinformed.

From this point of view, education and training should be directed toward helping the learner acquire and retain knowledge which is both correct and usable. The purpose of education is to prepare people to perform various activities of life more safely, more effectively and in more satisfying and comfortable ways than they would without the education. The goals of education and training programs should not only specify the content areas which should be learned (which they now do), but also should indicate some minimum level of usability. Indeed, it might be helpful to state the maximum level of misinformation, at least in some content areas, which is acceptable.

The multiple choice test has many advantages which include ease and objectivity of scoring, the ability to measure simple and complex knowledge in most content areas at most levels of knowledge, reliability and efficiency. For these reasons, it is expected that the multiple choice tests will be use more and more often. However, the knowledge of a person has many more characteristics or dimensions than is represented by the percentage correct score on a multiple choice test. Incorporating the concepts of usable knowledge and misinformation into testing produces test scores which are more representative of the way in which knowledge contributes to everyday decisions and actions in work, at home, and in play.

Self assessment testing

The observation that a person recognizes or recalls a correct response on a test does not allow any confident conclusion concerning whether the knowledge has been learned to a usable level (Figure 2). Similarly, if a person makes an incorrect response on a test, we do not know whether the person is uninformed or misinformed. The multiple choice Self Assessment Test provides remedies for both of these inadequacies. The Self Assessment Test was created for the primary purposes of (a) providing a more comprehensive (two-dimensional) measure of a person's knowledge and (b) the detection and identification of misinformation.

Over the last 10 years, the test has been altered with the aim of improving its ease and effectiveness of use by students for learning and by instructors for teaching and testing. For example, last year the paper-and-pencil answer sheet (Figure 3) was redesigned to simplify the instructions and to change the verbal labels of the five categories of self assessment to take advantage of Professor Gunnar Borg's (1986) extensive work on the psychophysics of physical exertion.

Also, last year an "Instructor's Summary" of the test results (Table 2) was added for teachers who prefer not to inspect the detailed printouts or who are intimidated by sets of numbers. In addition to giving summary statements of how well the group of students performed on the test, the Instructor's Summary lists those specific test items about which students as a group seem to be misinformed. On my tests in introductory psychology (with 50-200 students per class), 2-5% of the test items have a high percentage of Sure-but-Wrong responses indicating misinformation. This suggests that the Self Assessment Test may provide useful feedback to teachers-in-training.

Sometimes a test item is constructed in such a way that it misleads test takers. Once a test item with a high percentage of Sure-but-Wrong responses has been detected and identified, I have always been able to determine by simply reading the test item whether it is due to test taker misinformation or to a misleading construction of the test item. A printout of the complete item analysis (Table 3) is also provided.

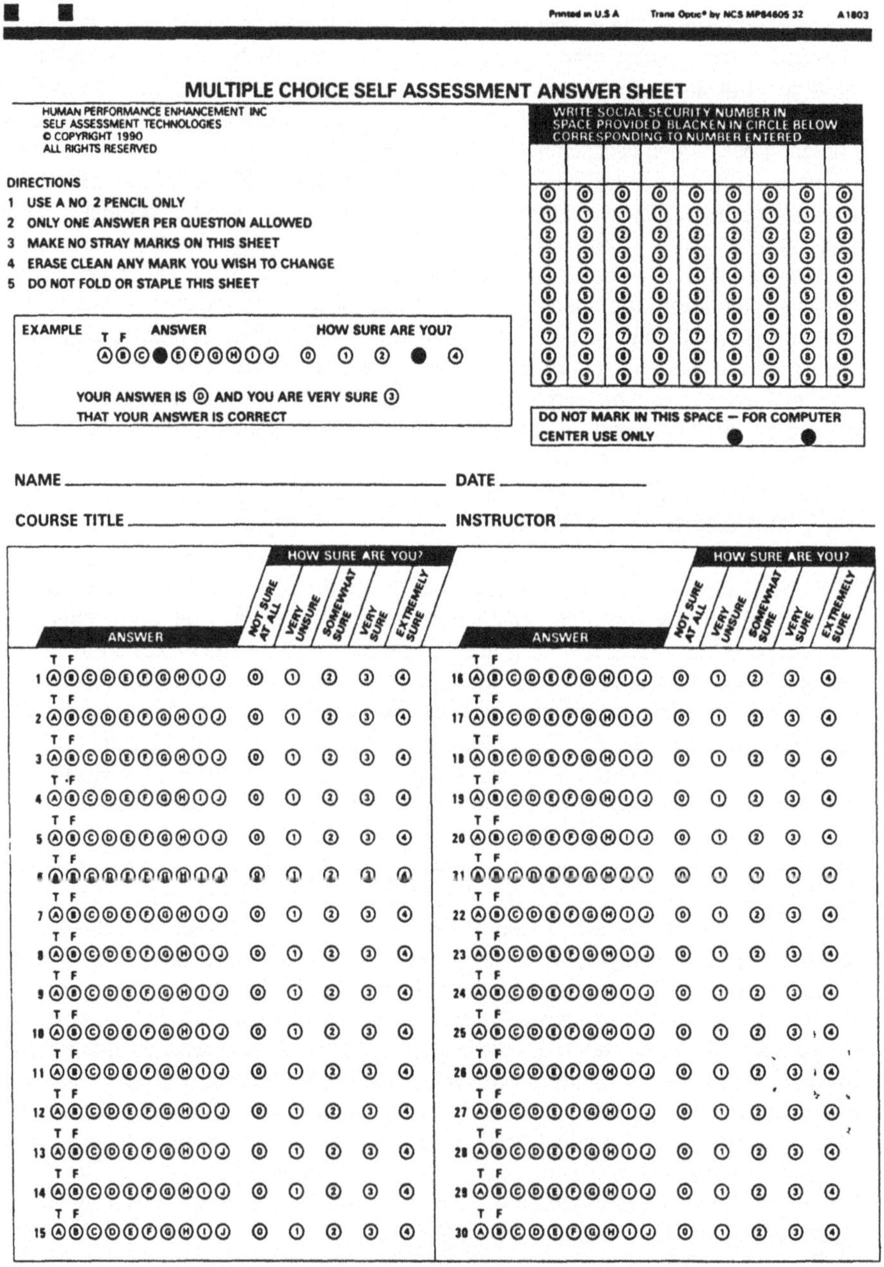

Printed in U.S.A   Trans Optic® by NCS MP54605 32   A1803

MULTIPLE CHOICE SELF ASSESSMENT ANSWER SHEET

HUMAN PERFORMANCE ENHANCEMENT INC
SELF ASSESSMENT TECHNOLOGIES
© COPYRIGHT 1990
ALL RIGHTS RESERVED

WRITE SOCIAL SECURITY NUMBER IN
SPACE PROVIDED BLACKEN IN CIRCLE BELOW
CORRESPONDING TO NUMBER ENTERED

DIRECTIONS
1 USE A NO 2 PENCIL ONLY
2 ONLY ONE ANSWER PER QUESTION ALLOWED
3 MAKE NO STRAY MARKS ON THIS SHEET
4 ERASE CLEAN ANY MARK YOU WISH TO CHANGE
5 DO NOT FOLD OR STAPLE THIS SHEET

EXAMPLE   T F   ANSWER   HOW SURE ARE YOU?

YOUR ANSWER IS Ⓓ AND YOU ARE VERY SURE ③
THAT YOUR ANSWER IS CORRECT

DO NOT MARK IN THIS SPACE — FOR COMPUTER
CENTER USE ONLY

NAME _____ DATE _____

COURSE TITLE _____ INSTRUCTOR _____

Figure 3. The multiple choice self assessment test answer sheet

Table 2. A summary of the test results which identifies for the instructor the test items about which the students are misinformed and uninformed.

Instructor's summary of test results

TEST: Psychol 201 Pretest   INSTRUCTOR: D.P.Hunt   DATE: 1 Aug 1990

OVERALL

A perfect score is a 100% Correct and a 100% Self Assessment (SA) Score. On this test : (a) the 75% CORRECT indicates that the test takers are somewhat knowledgeable on the topics of the test and (b) the 75% SA Score indicates that they are somewhat accurate in assessing their own knowledge.

The 46% SURE-and_CORRECT answers for the group indicates a somewhat low amount of usable knowledge. Of the correct answers, the test takers were sure of 61% of them, which indicates that they are inaccurate in the identification of the correct knowledge which they do possess.

TEST QUESTIONS RECOMMENDED FOR REVIEW

a. MISINFORMED. On 5% of their answers, the test takers were SURE that their answers were correct, but they were WRONG. This indicates that they possess a low amount of misinformation about the topics of the test. Test items which stand out because they have a relatively high percentage of SURE-but-WRONG answers are listed below.

Question Number	Percent Sure Wrong	Correct Answer	Most frequently chosen Sure-but-Wrong Answer
1	17	B	D
6	10	B	C
7	13	D	C
23	10	A	B
28	23	A	C

b. UNINFORMED. Questions which were answered correctly by fewer than 50% of the test takers are listed below.

Quest. Number	Correct Answer	Quest. Number	Correct Answer	Quest. Number	Correct Answer	Quest. Number	Correct Answer
4	D	27	D	40	B	50	C
11	C	29	A	41	C	51	B
13	D	30	D	43	A	53	A
15	D	31	B	44	A	55	A
16	B	32	C	45	D	56	C
22	C	33	B	46	B	57	B
24	D	38	C	47	A	58	C

TEST TAKERS WHO ARE ACCURATE IN SELF ASSESSMENTS

The test takers listed below were especially accurate in the self assessments of their own answers. They could be rewarded by, say, adding 3 percentage points to their % Correct Score.

ID Number	ID Number	ID Number
....58532	99938	26779
....58580	12464	58529
....52515	44574	58558

Table 3. An analysis of the questions and answers in the test.

Analysis of questions and answers

TEST: Psychol 201 Pretest   INSTRUCTOR: D.P.Hunt   DATE: 1 Aug 1990

QUEST.	% COR-RECT	% SA SCORE	SURE COR-RECT	SURE WRONG	UN-SURE COR-RECT	UN-SURE WRONG	NO SA RESP
1	17	49	3	17	13	67	0
2	73	74	7	0	67	27	0
3	63	64	10	7	53	30	0
4	23	72	0	3	23	73	0
5	97	84	57	0	40	3	0
6	50	60	20	10	30	40	0
7	33	54	13	13	20	53	0
8	83	75	15	0	67	17	0
9	57	67	7	0	50	43	0
10	63	72	10	0	53	37	0
11	23	66	0	3	23	73	0
12	50	65	7	3	43	43	3
13	13	59	0	7	13	77	3
14	60	68	10	0	47	40	3
15	43	57	3	7	40	47	3
16	37	73	10	0	23	63	3
17	100	75	30	0	67	0	3
18	67	73	10	0	57	30	3
19	77	79	20	0	53	23	3
20	77	80	30	0	43	23	3
21	80	71	23	7	50	13	7
22	47	75	7	3	37	50	3
23	27	57	7	10	17	63	3
24	47	76	10	0	33	53	3
25	60	76	13	0	47	37	3
26	57	76	23	3	33	37	3
27	23	75	0	3	23	70	3

28	10	44	0	23	10	63	3
29	13	84	0	3	13	77	7
30	7	81	0	3	3	87	7
31	37	64	0	3	37	53	7
32	30	74	0	0	30	63	7
33	40	70	0	3	37	53	7
.	.	.	.	.	.	.	.
.	.	.	.	.	.	.	.
.	.	.	.	.	.	.	.
44	7	89	0	0	3	90	7
45	23	59	0	3	20	73	3
46	30	75	0	0	30	63	7
47	10	67	0	0	10	83	7
48	67	76	7	0	57	33	3
49	67	73	10	0	53	33	3
50	20	78	3	0	13	77	7
51	30	71	0	0	30	67	3
52	53	61	0	0	53	40	7
53	47	79	10	0	33	50	7
54	50	69	0	3	47	43	7
55	33	68	0	0	33	60	7
56	33	72	0	3	33	57	7
57	17	69	0	7	17	67	10
58	43	71	3	3	40	47	7

A conclusion from "human self assessment process theory" is that the quality of people's performance depends both on the knowledge they possess and the confidence with which they possess it. Thus, in using the Self Assessment Test, it seems desirable to reward those test takers who are most accurate in assessing their own knowledge, i.e., developing the skill of self assessment is worth expending effort and should be rewarded. To provide such a reward, and to provide an incentive for engaging in self assessment, the percentage Correct score can be increased 3% for the accurate self assessors.

This raises questions concerning the value of various levels of sureness, e.g., usable knowledge. I have calculated a % SA Score (Table 4), but have given little systematic thought to identifying the factors which should be considered in establishing the relative weightings of the % Correct and the % SA Score.

Table 4. A listing of the test takers and their Percentage Correct, Percentage Self Assessment Score, and Percentage of Sure-and-Correct, Unsure-and-Correct, Unsure-and-Wrong, and Sure-but-Wrong answers. An asterisk (*) indicates that the test taker was one of the more accurate in assessing the correctness of their own answers.

STUDENTS' LISTING

TEST: Psychol 201 Pretest   INSTRUCTOR: D.P.Hunt   DATE: 1 Aug 1990

ID NUMBER	% CORR	% SA SCORE	SURE CORR	SURE WRONG	UN-SURE CORR	UN-SURE WRONG	NOSA RESP
.89415	63	67	24	3	38	34	0
.51696	71	75	50	7	22	20	0
.58543	89	86	88	7	2	4	0
.58532*	61	69	14	4	46	36	0
.52535	71	75	41	7	30	21	0
.58580*	82	83	52	0	30	18	0
.52515*	38	61	25	25	13	38	0
.99938	95	92	91	4	4	2	0
.52531	70	79	52	13	18	18	0
.99958*	89	84	89	11	0	0	0
.58592	86	82	30	0	55	14	0
.52511	79	77	38	4	41	18	0
.12464*	54	71	21	2	32	45	0
.44574*	95	92	80	2	14	4	0
.58521	88	85	75	5	13	7	0
.58531	91	86	88	9	4	0	0
.52531	75	67	68	20	7	5	0
.52525	48	46	45	38	4	14	0
.52553	88	83	30	0	57	13	0
.52527	57	60	41	16	14	27	2
.26779*	95	91	91	4	4	2	0
.52539	55	61	46	18	9	27	0
.52529	79	79	45	4	34	18	0
.50074	68	73	38	5	30	27	0
.58529*	55	68	18	5	36	39	2
.58558*	82	85	66	4	16	14	0
.58580	70	65	50	14	20	16	0
Highest	95	92	91	38	57	45	2
Median	75	75	46	5	18	18	0
Lowest	38	46	14	0	0	0	0

The computer analysis of the answer sheets of the Self Assessment Test provides a printout of the scores (Table 4) which can be displayed for knowledge of results to the test takers. The test takers who are more accurate in their self assessment and, thus, most deserving of the test bonus, are identified with an asterisk (*). A printout of the answers and self assessment responses of each test taker on each test item is also available (Table 5).

Table 5. The responses of a test taker on each test item.

RESULTS OF A PERSON   ID 89415

TEST: Psychol 201 Pretest   INSTRUCTOR: D.P.Hunt   DATE: 1 Aug 1990

Ans: ccbdb bbddd acadd bcbaa dbcdb bdbcb bcabd bcdbb dcdcd cdccc bdacb dac
SA: 33234 43444 31333 44113 42303 32421 03434 33340 23303 03443 33433 234
C/W: wccc cwccc wcwcc ccccc cwwcc ccwww wcwcc wcwcc wcwwc wwccc cccw wwc

   The number of correct answers was 37 out of 58
                     ... so the percentage CORRECT was 63%
                     ... and SELF-ASSESSMENT Score was 67%

   The percentage of:
                     SURE-and-CORRECT answers was 24%
                     SURE-but-WRONG answers was 3%
           UNSURE-but-CORRECT answers was 38%
           UNSURE-and-WRONG answers was 34%
                     No SA Responses was 0%

Other features of self assessment testing

Motivational effects. When I first began employing multiple choice self assessment tests in my classes in 1983, I was concerned about whether its use might have some detrimental side effects which I had not anticipated. To detect any such unanticipated effects and, also, to obtain data about the reactions of students to the tests, at the end of the 3-4 month course of study I administered an open-ended questionnaire in which the students were asked to list and comment about the advantages and disadvantages as they perceived them of the self assessment tests for learning, grading, testing, and teaching.

There were diverse reactions, but a surprise was that about 40% of the students indicated that they study more to prepare for the Self Assessment Test, e.g., "to be able to mark that I am sure of my answer," than they do for the usual multiple choice test.

The possibility that an instructor can increase the time spent by students in studying the material simply by employing the Self Assessment Test, is attractive and should be pursued further.

Reduction of gender bias. Multiple choice tests are widely employed in the United States to help decide who will receive scholarships, who will be admitted to educational institutions, who will be given a license to practice medicine, law, psychology, etc. Critics of such tests argue that multiple choice tests are biased against various groups of people, such as females. To be biased means, here, that if a male and female know the same amount about the topics of the test, then one of them will obtain a lower score on the test than the other.

For example, about 60% of the National Merit Scholars, in which the semifinalists are selected largely based on the results of a multiple choice Scholastic Aptitude Test, are boys.

Our own unpublished research findings (Table 6) found that the difference in the percentage correct answers between male and female university students was reduced when the Self Assessment Test was used compared to the usual multiple choice test.

Table 6. The mean number of correct answers (on 50-item mathematics and verbal test) using the usual multiple choice test and using the Self Assessment Test. The total number of subjects was 120; 30 males and 30 females was administered each of the two tests.

Kinds of Multiple Choice Test	Gender of Test Takers	Number Correct	Standard Deviation
Usual multiple choice test	Female	23.9	0.54
	Male	29.2	0.67
Self Assessment Test	Female	27.8	0.53
	Male	29.7	0.55

The reasons that Self Assessment Testing may affect the test score differences are not clear. It might be that allowing the test takers to express their doubt or certainty about the correctness of their answers helps to reduce the anxiety of the test taker; and two people might bring different levels of anxiety to the test taking situation. Another possibility is that there may be some factor in the differences between the sexes, e.g., in the way in which boys and girls are reared, which produces different kinds of risk taking attitudes or different processing of information under risky situations. For example, females might rely more on recall than do males, and asking them to assess the correctness of their answer might produce a more careful review of their answers.

Summary of benefits of Self Assessment Testing. In summary, the potential benefits of self assessment as I perceive them at this time are:

1. Obtain a more comprehensive measure of a person's usable knowledge.
2. Detect areas of knowledge in which a person is misinformed.
3. Identify test items which may be misleadingly constructed.
4. Encourage more effective study.
5. Make people aware of self assessment as an important part of their performance.
6. Provide practice with feedback in making self assessments.
7. Enhance learning for some topics for some people.
8. Make testing a learning experience.
9. Make testing and learning less anxiety producing, more enjoyable and satisfying.

References

Borg, G. (1986). Perception of exertion in physical exercise. London, England: Macmillan

Hassmen, P. & Hunt, D.P. (1992, unpublished). Human self assessment in multiple choice testing

Hunt, D.P. and Furustig, H. (1989). Being informed, being misinformed and disinformation: A human learning and decision making approach. Technical Report PM 56:238, 1989-04-21, Karlstad: Institution 56 Manniska Maskin System

Hunt, D.P. (1982). Effects of human self-assessment responding on learning. Journal of Applied Psychology, 67, 75-82

Hunt, D.P. (1991). Self assessment technology: Multiple choice self assessment testing. Human Performance Enhancement, Inc. Report, 12 pp.

Sams, M.R. (1989). Effects of observational assessments and patterns of success-failure on self-confidence. Unpublished Dissertation. New Mexico State University, Las Cruces, New Mexico

"The Master said, Yu, shall I teach you what knowledge is?"

"When you know a thing, to recognize that you know it, and when you do not know a thing, to recognize that you do not know it."

"That is knowledge."

Confucius (551-479 B.C.)

Using Testing to Provide Feedback to Support Instruction: A Reexamination of the Role of Assessment in Educational Organizations

James E. Bruno

UCLA, Graduate School of Education, 131 Moore Hall, Los Angeles, California 90024, U.S.A.

Abstract: Presently used assessment practices mainly focus on *summative* evaluation and ignore the important function of *formative* evaluation. Formative evaluation is needed to provide information and feedback to the learner as well as to teachers and school administrators in the instructional program. This paper demonstrates how a technology based Information Referenced Testing (IRT) concept can be used to provide formative evaluation feedback to the instructional leadership *triad* of learner, instructor and educational administrator. Illustration of various IRT generated feedback reports and a survey of applications to educational organizations are provided.

1 Introduction

There are three major types of policy issues and decision making options that are associated with student testing and assessment. These are the *why* of assessment, the *how* of assessment and the *what* of assessment.

Why societies assess their young is embedded in the "collective" mindset of mankind or, as Carl Jung would say, the "collective unconscious" of man. The rite of passage of youth, the Bar Mitzvah, the confirmation, the SAT, etc., always seems to entail some sort of ordeal or rite of passage or performance based assessment for the youth of a culture. Even the ancient Greeks were keenly aware of the need for assessment. When the Sphinx asked Oedipus "what walks with four legs then two legs then three legs" and Oedipus answered "man", he was judged by the Sphinx as being "king" material. History's first Norm Referenced Test (certainly a test item with high discriminant validity) was thus recorded. Over the millennia of man, only the format of testing -- oral, performance, written-essay, and multiple choice has changed. The purpose of testing and assessment, however, seems to remain the same regardless of culture -- sorting of individuals. It is only recently that testing to support instructional programs in educational organizations and provide feedback to the instructional leadership triad -- students (and parents), teachers and educational administrators -- has come into greater prominence.

2 Why We Test

In a modern, twentieth century, industrial state, the *"why"* of testing in the classroom serves two important functions. These are:

(1) *Summative Evaluation* -- judgment, selection and sorting of students
(2) *Formative Evaluation* -- assessment of a student to promote classroom learning to make the instructional process more efficient, and to support instructional programs with detailed feedback.

Tests, therefore, can be used in the classroom for both summative and formative evaluation purposes. Yet most school organizations emphasize the summative over the formative nature of testing and assessment. Naturally it is the latter purpose of testing and assessment, i. e. *formative* evaluation, that is perhaps most important for promoting classroom learning and providing appropriate feedback to support instructional programs. Assessments of students have to be concerned about the natural tendency of students to *forget* material (previously taught) and the propensity of students to *learn new material* (to be taught). Both *forgetting* old information and *learning* new information are strongly related to the expressed confidence or "fluency" that students have in their current information base -- i.e. their levels of reliable (accurate and confident) information. See Figure 1 for a hypothetical depiction of how learning new information (i.e. percents) is based on the confidence or "fluency" in previously learned information (i.e. decimals). Also note how previously learned information is forgotten over time in direct relationship to expressed confidence in that information.

The research of Hunt (1982) nicely demonstrated how confidence in information can impact on student learning rates (by as much as 20%). Hunt (1977) refers to the self-assessment process as a mechanism for the enhancement of learning based on building fluency or confidence in information.

Recent empirical research by Klentschy (1992 a, b) and Baxter (1993) has largely confirmed Hunt's research with regard to classroom instruction in mathematics. As depicted in Figure 1 this research supports the notion that if one can build student confidence and fluency in information (informed), not only will this confidence be sustaining, but it can also lead to an acceleration in learning new information to high levels of confidence Klentschy's research (1992), in particular, examines this hypothesis in a pre-post and post-post test type of experimental design. In his study, specific concepts in mathematics were examined for both their decay over time and in their precursive relationship to the learning of other mathematics concepts.

Each "party" to the instructional process requires different attributes of the assessment process. "Why" we assess students, at least from the *classroom teacher perspective*, is basically twofold.

(1) Judgement, selection and sorting (grades and placement) -- *Summative Evaluation*
(2) Supporting instruction via detailed feedback. Ultimately assessment is needed to enhance learning of new information and minimize forgetting of old information -- *Formative Evaluation*

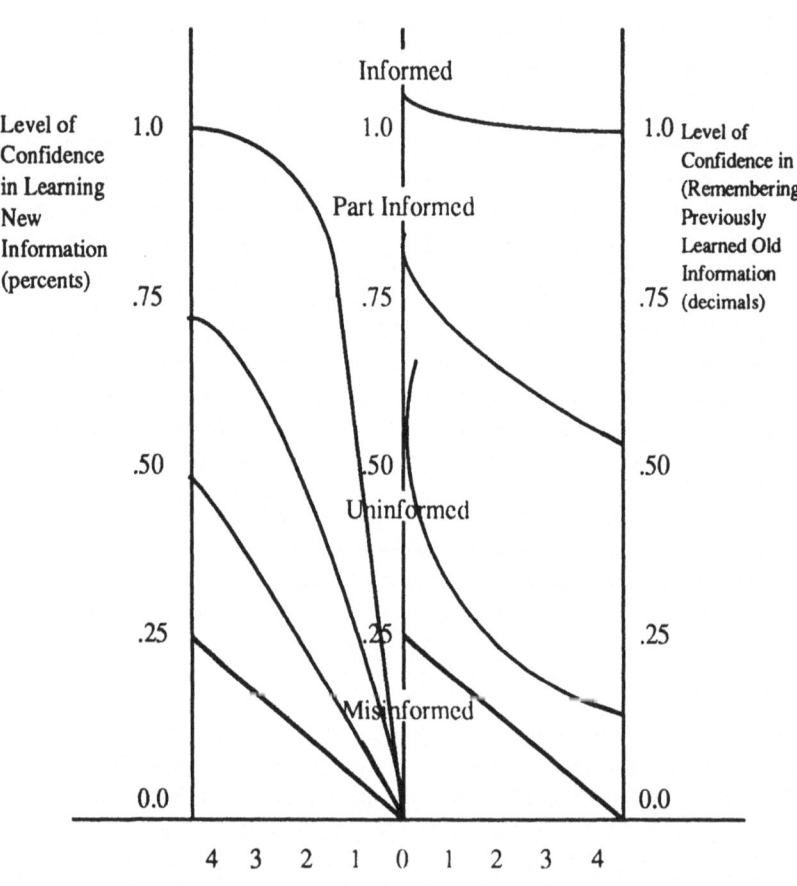

Figure 1
Hypothetical Relationship of
Learning New Information and
Forgetting Previously Learned Information as a
Function of Expressed *Confidence* in Information

From the *student (and parent) perspective* the "why" of testing and assessment includes:

(1)   Ensuring that the student is "on line" or "on schedule" in terms of meeting the educational objectives of the school, family or culture (an early warning system)

(2)   Articulating the school's instructional program directly to students (and parents) so that students can help themselves learn and shoulder some of the burden of instruction (an individually prescribed plan of action).

But there are two other recipients of assessment information, the school site or instructional unit educational administrator and the central office or the across instructional unit educational administrator. From the *school administrative perspective*, at the school site level, the "why" of assessment should include:

(1)   Resource allocation (micro level) to promote the overall effectiveness of the instructional program at the school site or instructional unit level

(2)   Possible use of assessment information to provide teacher accountability, assist in teacher assignment to classes, and design staff inservice programs

(3)   Curriculum alignment and planning to support the instructional objectives of the school site or instructional unit

From the across unit or central office or *state perspective*, the "why" of testing might include:

(1)   Overall educational accountability for the system

(2)   Resource allocation (macro level) to promote the central office or the state's instructional objectives.

All of these parties to the assessment process -- teacher, student (and parent), instructional unit administration and the central or state (across instructional units) administration -- have a stake in making the assessment process more efficient and effective. In order to service these differing needs, assessment has to be more *formative* in nature than summative and provide detailed feedback to students, teachers and administrators to support instruction. Most important, this feedback has to be placed in a format to optimize its use in decision making and policy formulation.

3 How We Test

The *"how"* we assess students is based on two major considerations with regard to how student responses are evaluated. The evaluation of student responses can either be *objectively* scored (multiple choice, low cost, machine scoring) or *subjectively* scored (performance based, portfolio analysis, teacher evaluation). There are many important reasons why *objectively* scored testing is presently favored over *subjectively* scored testing. Some of these reasons include and are based on *economic* (cost and economic viability), legal (bias and due process), *feedback* (consistency, report generation across evaluations) and *educational* (domain of skills, information for instructional decision making) considerations. In general objective testing is preferred over subjective assessment because it (1) can assess a wider domain of educational concepts and skills (2) eliminates evaluator bias from the assessment process (3) can be used to provide consistent feedback at different levels of aggregation (student, class, school, district, and state) and (4) most importantly, is economically viable for the educational agency to perform on a mass scale since assessment procedures can be made optically scannable.

Presently there are *three* categories of objectively scored tests that are available to educational agencies. These are the *Norm* Referenced Test (NRT), the *Criterion* Referenced Test (CRT), and the *Information* Referenced Test (IRT). The following table depicts some of the main characteristics of each type of test.

Type	*Mechanics*	*Type of Objective Scoring*
Norm Referenced Test	Student Score is Referenced to a Normal Distribution Curve (percentile)	Right or Wrong (one-dimensional and non-reproducible)
Criterion Referenced Test	Student Score is Referenced to a Selection Standard Score (percent correct)	Right or Wrong Test (one-dimensional and non-reproducible)
Information Referenced Test	Student Score is Referenced to Information Standard of attainment (percent of full information)	MCW-APM (Modified Confidence Weighted- Admissible Probability Measurement) (two-dimensional and reproducible)

Note that these three types of assessment formats differ with regard to both the *dimensionality* of the scoring system (one or two) and the *standard* upon which to evaluate student attainment (each other, percent correct, information).

The *Norm* Referenced Test such as the Stanford Achievement Test (SAT), and the Comprehensive Test of Basic Skills (CTBS), is typically used by educational agencies (school districts, government agencies, universities) for *summative* evaluation. Because of the one-dimensional nature of the scoring system (recognition only) they are notoriously sensitive to student guessing, test preparation and culture bias. In addition, the important pedagogical consideration of misinformation cannot be assessed with one-dimensional scoring systems. These limitations make them nearly useless sources of formative evaluation and feedback information for the teacher (as well as students and administrative personnel) for promoting classroom learning and supporting instruction. The fact that the scoring system used on a Norm and Criterion referenced tests are *not reproducible*, since guessing is implicitly encouraged, severely limits its practical use in classroom formative evaluation. The one-dimensional nature of the scoring system also produces "overestimates" of attainment and subject matter mastery, especially for low and middle attaining students. Finally, misinformation and the lack of information cannot be assessed, in spite of the fact that both of these information states required differing instructional sequences -- re-education and instruction -- by the classroom teacher.

There is an important scientific tautology, that should be noted, with regard to the use of Norm Referenced Tests to support instruction. First, there is an assumption made that there is a *normal* distribution model of academic attainment and that test items merely have to be designed to fit the model (discriminant validity). Rather than a model first being derived from the facts (items) as in most scientific endeavors, here the facts are aligned to the model via the discriminant validity of the test item. In short, the Norm Referenced Test is generally considered by teachers to be nearly useless for measuring anything but the narrowest, and most times, instructionally irrelevant set of skills (trick questions, riddles). While the results of Norm Referenced Tests seem valid for the highest and lowest attaining students, the great majority of students in the middle to low ranges are ill served by the information derived from these assessments.

The *Criterion* Referenced Test (CRT), such as a mathematics or language placement test, is far more "in line" with teacher needs for feedback to improve instructional decision making and classroom learning. Unfortunately, due to the one-dimensional nature of the system of scoring or evaluating (Right or Wrong) student responses on CRT's, they are also extremely sensitive to guessing and tend to yield over estimates of actual attainment -- especially in the middle and low ranges. In addition there is a strong inverse relationship between the selection standard or criterion score used on a CRT and the number of false positives and false negatives generated by the assessment process. The dual problems of determining a selection standard and the one-dimensional nature of the scoring system, make CRT's extremely limited for supporting instructional programs.

The *Information* Referenced Test is a hybrid NRT and CRT. Like a CRT, the main emphasis is to align the assessment process to instructional objectives in order to support classroom instruction. Like an NRT, the IRT score is referenced to a fixed standard or model of attainment. The Information Referenced model of attainment is based on information theory since student responses are scored using a two-dimensional scoring system (recognition and confidence). Evaluating students on a standard of being informed, part informed, uninformed and misinformed is more pedagogically aligned (review, instruction, re-education) to teacher needs and instructional support than either a percent correct score (CRT) of a percentile score (NRT). The scoring system used with an IRT format is not only two-dimensional (recognition plus confidence) in nature, but is also *reproducible*. With a *reproducible* test scoring system, the maximum score for the student is obtained *if and only if* the student does not guess, but answers each test item on the assessment with his or her actual information, i. e., does not overvalue information. The IRT procedure is fully *optically scannable* (see Figure 2) and generates an *individual education plan* (IEP) for purposes of formative evaluation for each student assessed (see Figure 3). A *class information needs profile* (CINP) (see Figure 4) is provided for the teacher and a *school information needs profile* (SINP) (see Figure 5) is also provided for the school administrator. In short, feedback with an IRT procedure is used to identify the *specific* information needs (misinformation, partial information, no information) of each

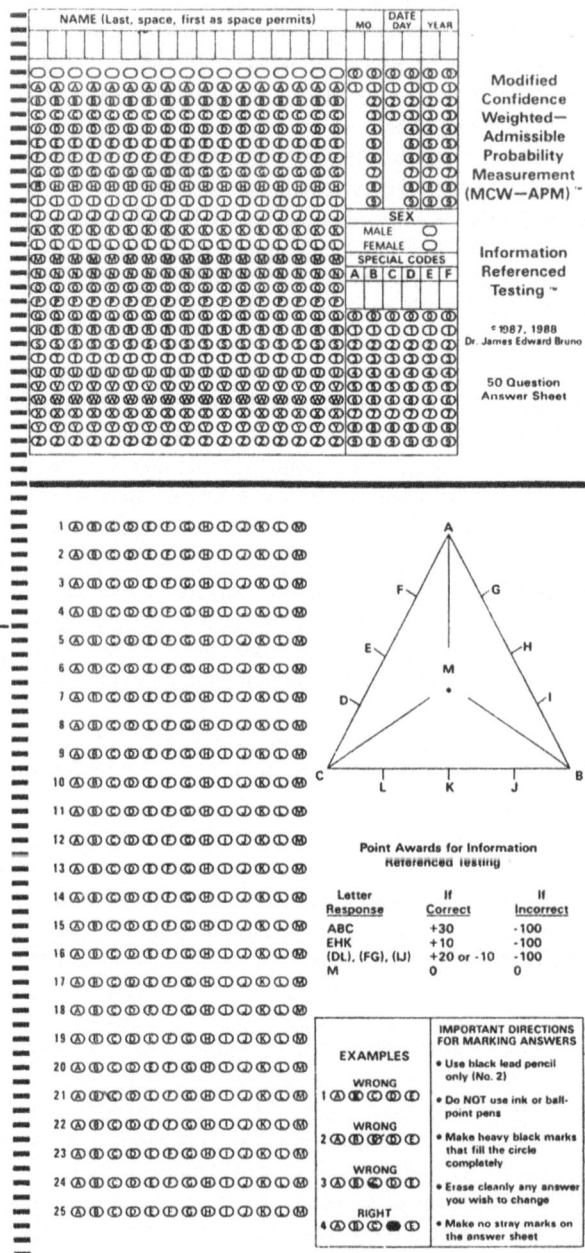

Figure 2
Optical Scan Sheet to Support
Two-Dimensional Information Referenced Testing (IRT)
in the Classroom
(simplified 7 response option and triangle 14 response option)

NAME (Last, First, MI.)   DATE

MO. DAY YR.

SEX   Male ○
      Female ○

SPECIAL CODES
A B C D E F

MODIFIED
MODIFIED CONFIDENCE WEIGHTED—
ADMISSIBLE PROBABILITY MEASUREMENT
(MMCW—APM) ™

Information Referenced Testing ™

© 1988 Dr. James Edward Bruno

30 QUESTION ANSWER SHEET

I AM SURE			I AM NOT SURE			I DON'T KNOW
1. Ⓐ	Ⓑ	Ⓒ	ⒶⒷ	ⒷⒸ	ⒶⒸ	?
2. Ⓐ	Ⓑ	Ⓒ	ⒶⒷ	ⒷⒸ	ⒶⒸ	?
3. Ⓐ	Ⓑ	Ⓒ	ⒶⒷ	ⒷⒸ	ⒶⒸ	?
4. Ⓐ	Ⓑ	Ⓒ	ⒶⒷ	ⒷⒸ	ⒶⒸ	?
5. Ⓐ	Ⓑ	Ⓒ	ⒶⒷ	ⒷⒸ	ⒶⒸ	?
6. Ⓐ	Ⓑ	Ⓒ	ⒶⒷ	ⒷⒸ	ⒶⒸ	?
7. Ⓐ	Ⓑ	Ⓒ	ⒶⒷ	ⒷⒸ	ⒶⒸ	?
8. Ⓐ	Ⓑ	Ⓒ	ⒶⒷ	ⒷⒸ	ⒶⒸ	?
9. Ⓐ	Ⓑ	Ⓒ	ⒶⒷ	ⒷⒸ	ⒶⒸ	?
10. Ⓐ	Ⓑ	Ⓒ	ⒶⒷ	ⒷⒸ	ⒶⒸ	?
11. Ⓐ	Ⓑ	Ⓒ	ⒶⒷ	ⒷⒸ	ⒶⒸ	?
12. Ⓐ	Ⓑ	Ⓒ	ⒶⒷ	ⒷⒸ	ⒶⒸ	?
13. Ⓐ	Ⓑ	Ⓒ	ⒶⒷ	ⒷⒸ	ⒶⒸ	?
14. Ⓐ	Ⓑ	Ⓒ	ⒶⒷ	ⒷⒸ	ⒶⒸ	?
15. Ⓐ	Ⓑ	Ⓒ	ⒶⒷ	ⒷⒸ	ⒶⒸ	?

IMPORTANT
DIRECTIONS FOR MARKING ANSWERS

USE NO. 2 PENCIL ONLY

- Do NOT use ink or ballpoint pens.
- Make heavy black marks that fill the circle completely. See example below.

I AM SURE	I AM NOT SURE	I DON'T KNOW
Ⓐ ● Ⓒ	ⒶⒷ ⒷⒸ ⒶⒸ	?

- Erase cleanly any answer you wish to change.
- Make no stray marks on the answer sheet.

INFORMATION REFERENCED TESTING ™
RESPONSE OPTIONS AND POINT AWARDS

Letter Response	If Correct	If Incorrect
I AM SURE Ⓐ Ⓑ Ⓒ	+30	-100
I AM NOT SURE Either ⒶⒷ Either ⒷⒸ Either ⒶⒸ	+10	-100
I DON'T KNOW ?	0	0

ⒶⒷ Between Choices A and B

ⒷⒸ Between Choices B and C

ⒶⒸ Between Choices A and C

Figure 2
(continued)

IEP -- CCN

EXAMINEE INDIVIDUAL EDUCATION PLAN (IEP)

EXAMINEE NAME
EXAM NAME   CRITICAL CARE ASSESSMENT TOOL
EXAM CODE   1
SCHOOL NAME
SCHOOL SITE CODE   1
INSTRUCTOR NAME
NUMBER OF QUESTIONS 50
PROCESSING CODE (A=MCW-APM B=MCW-APM AND RW) =B

FORMATIVE EVALUATION

EXAMINEE MISINFORMATION ON EXAMINATION

CONCEPTS WHERE YOU WERE SURE OF AN ANSWER BUT WERE WRONG

HAVE INSTRUCTOR EXPLAIN WHY THE ANSWER YOU THOUGHT WAS CORRECT WAS WRONG AND WHY
ANOTHER ANSWER WAS CORRECT

TEST ITEM (INF STATE)   DESCRIPTION - INSTRUCTIONAL CROSS REFERENCE
    12   M   FLUID & ELECTROLYTE CHANGES IN DIURETIC THERAPY
                                 KEE FLUID & ELECTROLYTES WITH CLINICAL APPLICATIONS, 3 ED.
    13   M   CARDIAC IRRITABILITY IN HYPERCALCEMIA
                                 THE NURSES GUIDE TO F/ELECTROLYTE BALANCE 2ND ED.
     •   •
     •   •

EXAMINEE UNINFORMED (LACKS INFORMATION) RESPONSES

CONCEPTS THAT YOU SAID YOU DIDNT KNOW - HAVE YOUR INSTRUCTOR EXPLAIN THESE CONCEPTS TO
YOU

TEST ITEM (INF STATE)   DESCRIPTION - INSTRUCTIONAL CROSS REFERENCE
    39   D   FINDINGS IN NORMAL PRESSURE HYDROCEPHALUS
                                 CARINI, NEUROLOGICAL AND NEUROSURGICAL NSG. 1979 P. 241

EXAMINEE PARTIALLY INFORMED ITEMS ON EXAMINATION

CONCEPTS WHERE YOU WERENT SURE OF THE ANSWER - HAVE YOUR INSTRUCTOR REVIEW THESE
CONCEPTS WITH YOU

TEST ITEM (INF STATE)   DESCRIPTION INSTRUCTIONAL CROSS REFERENCE

EXAMINEE FULLY INFORMED CONCEPTS (RELIABLE INFORMATION)

CONCEPTS THAT YOU SAID YOU WERE SURE OF THE ANSWER AND THAT ANSWER WAS CORRECT -
YOU HAVE RELIABLE INFORMATION IN THESE AREAS - KEEP UP THE GOOD WORK

TEST ITEM (INF STATE)   DESCRIPTION - INSTRUCTIONAL CROSS REFERENCE
    1   I   PATHO PHYSIOLOGY OF SHOCK
                                 PERRY SHOCK MOSBY, 1983
    2   I   CLIN. PRESENTATION OF SHOCK
                                 PARK, CARDIAC SHOCK CCQ 2; 43-53
     •   •
     •   •

Figure 2
(continued)

STUDENT COGNITIVE MAP

PERCENT INFORMED	0.92
PERCENT UNINFORMED	0.02
PERCENT PART INFORMED	0.0
PERCENT MISINFORMED	0.06
PERCENT RIGHT WITH RW	0.94
PERCENT WRONG WITH RW	0.06

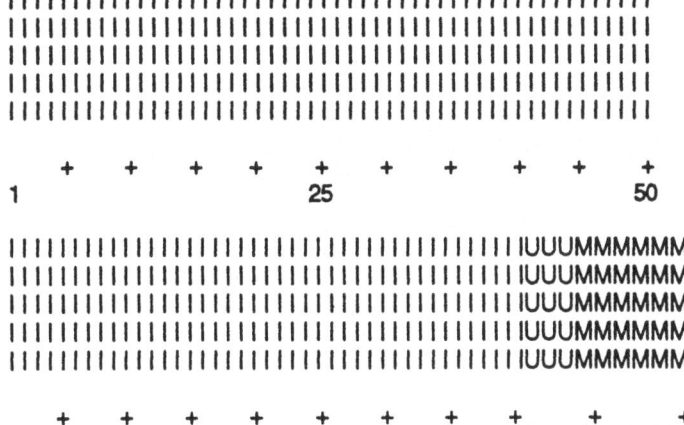

Figure 3
Feedback #1a: Excellent Information
Student Individual Education Plan (IEP) or Audit
Generated with the IRT Procedure
(audits of two critical care nurses with differing information needs)

IEP – CCN

EXAMINEE INDIVIDUAL EDUCATION PLAN (IEP)

EXAMINEE NAME
EXAM NAME   CRITICAL CARE ASSESSMENT TOOL
EXAM CODE   1
SCHOOL NAME
SCHOOL SITE CODE   1
INSTRUCTOR NAME
NUMBER OF QUESTIONS 50
PROCESSING CODE (A=MCW-APM B=MCW-APM AND RW) =B

FORMATIVE EVALUATION

EXAMINEE MISINFORMATION ON EXAMINATION

CONCEPTS WHERE YOU WERE SURE OF AN ANSWER BUT WERE WRONG

HAVE INSTRUCTOR EXPLAIN WHY THE ANSWER YOU THOUGHT WAS CORRECT WAS WRONG AND WHY
ANOTHER ANSWER WAS CORRECT

TEST ITEM (INF STATE)   DESCRIPTION - INSTRUCTIONAL CROSS REFERENCE
     1   M   PATHO PHYSIOLOGY OF SHOCK
                               PERRY SHOCK MOSBY, 1983
     2   M   CLIN. PRESENTATION OF SHOCK
                               PARK, CARDIAC SHOCK CCQ 2; 43-53
        •   •
        •   •

EXAMINEE UNINFORMED (LACKS INFORMATION) RESPONSES

CONCEPTS THAT YOU SAID YOU DIDNT KNOW -
HAVE YOUR INSTRUCTOR EXPLAIN THESE CONCEPTS TO YOU

TEST ITEM (INF STATE)   DESCRIPTION - INSTRUCTIONAL CROSS REFERENCE
     7   U   DECREASED GFR AND ELECTROLYTE CHANGES
                               STARK, J. RENAL FAILURE: IMBALANCES INEVITABLE CCN DEC 198
     9   U   EKG CHANGES WITH POTASSIUM CHANGES
                               KAPPAGODA, ARRYTHMIA CASE STUDIES P. 185
        •   •
        •   •

EXAMINEE PARTIALLY INFORMED ITEMS ON EXAMINATION

CONCEPTS WHERE YOU WERENT SURE OF THE ANSWER -
HAVE YOUR INSTRUCTOR REVIEW THESE CONCEPTS WITH YOU

TEST ITEM (INF STATE)   DESCRIPTION INSTRUCTIONAL CROSS REFERENCE
    11   P   EKG CHANGES WITH DIG TOXICITY
                               WINKLE, R. CARDIAC ARRYTHMIA CURRENT DX & MGT, 1983
    18   P   APPROPRIATE BLOOK PRODUCTS REPLACEMENT
                               DAROVIC CCN 2:6 P. 36-46, 1982
        •   •
        •   •

Figure 3
Feedback #1a
(continued)

EXAMINEE FULLY INFORMED CONCEPTS (RELIABLE INFORMATION)

CONCEPTS THAT YOU SAID YOU WERE SURE OF THE ANSWER AND THAT ANSWER WAS CORRECT -
YOU HAVE RELIABLE INFORMATION IN THESE AREAS - KEEP UP THE GOOD WORK

TEST ITEM (INF STATE)		DESCRIPTION - INSTRUCTIONAL CROSS REFERENCE
3	I	PHARMACOLOGIC RX OF CARD. SHOCK
		BARROWS SHOCK DEMANDS DRUGS NSG. 82 PP. 34-40
4	I	PACEMAKER TYPES
		SANDERSON & <KLURTH THE CARDIAC PT. SAUNDERS, 1983
.	.	
.	.	

STUDENT COGNITIVE MAP

PERCENT INFORMED	0.16
PERCENT UNINFORMED	0.36
PERCENT PART INFORMED	0.24
PERCENT MISINFORMED	0.22
PERCENT RIGHT WITH RW	0.60
PERCENT WRONG WITH RW	0.34

```
IIIIIIIIIIIIIIII PPPPPPPPPPPPPPPPPPPPPPPPPPPUUUUUUUUUUUU
IIIIIIIIIIIIIIII PPPPPPPPPPPPPPPPPPPPPPPPPPPUUUUUUUUUUUU
IIIIIIIIIIIIIIII PPPPPPPPPPPPPPPPPPPPPPPPPPPUUUUUUUUUUUU
IIIIIIIIIIIIIIII PPPPPPPPPPPPPPPPPPPPPPPPPPPUUUUUUUUUUUU
IIIIIIIIIIIIIIII PPPPPPPPPPPPPPPPPPPPPPPPPPPUUUUUUUUUUUU
   +    +    +    +         +    +    +    +    +    +
   1                       25                       50

UUUUUUUUUUUUUUUUUUUUUUUUUUUUUUUUMMMMMMMMMMMMMMMMMMMMMMMM
UUUUUUUUUUUUUUUUUUUUUUUUUUUUUUUUMMMMMMMMMMMMMMMMMMMMMMMM
UUUUUUUUUUUUUUUUUUUUUUUUUUUUUUUUMMMMMMMMMMMMMMMMMMMMMMMM
UUUUUUUUUUUUUUUUUUUUUUUUUUUUUUUUMMMMMMMMMMMMMMMMMMMMMMMM
UUUUUUUUUUUUUUUUUUUUUUUUUUUUUUUUMMMMMMMMMMMMMMMMMMMMMMMM
   +    +    +    +    +    +    +    +    +    +
  55   60   65   70   75   80   85   90   95  100
```

Figure 3
Feedback #1b: Poor Information
Student Individual Education Plan (IEP) or Audit
Generated with the IRT Procedure
(audits of two critical care nurses with differing information needs)

CINP

MISINFORMED EXAMINATION ITEMS FOR YOUR EXAMINEES
YOUR EXAMINEES HAVE WRONG INFORMATION IN THESE CONCEPT AREAS - MISINFORMED

HAVE WORKSHOP COORDINATOR DEVELOP INSTRUCTIONAL MATERIALS
DEMONSTRATE MISCONCEPTIONS - FOLLOW THIS WITH ACCURATE INFORMATION

ITEM	NUMBER	PERCENT	CONCEPT DESCRIPTION/CROSS REFERENCE
14	14.00	0.64	FIND AREA - LENGTH TIMES WIDTH
			ADDISION WESSLEY 4TH GRADE CHAPTER 6 P 144-145
3	8.00	0.36	WRITE DECIMALS TO HUNDREDTHS PLACE
			ADDISON WESSLEY 4TH GRADE CHAPTER 15 P 344-345
•	•	•	•
•	•	•	•

UNINFORMED TEST ITEMS ON THE EXAMINATION
EXAMINEES GENERALLY LACKINFORMATION IN THESE CONCEPT AREAS

BASIC INSTRUCTION NEEDED

HAVE WORKSHOP COORDINATOR PREPARE INSTRUCTIONAL MATERIALS TO TEACH THESE BASIC
CONCEPTS

ITEM	NUMBER	PERCENT	CONCEPT DESCRIPTION/CROSS REFERENCE
28	12.00	0.55	MAP TO SOLVE PROBLEMS (DECIMALS)
			ADDISON WESSLEY 4TH GRADE CHAPTER 14 P 354-355
4	10.00	0.45	ROMAN NUMERALS
			ADDISON WESSLEY 4TH GRADE CHAPTER 2 P 37
•	•	•	•
•	•	•	•

PART INFORMED CONCEPT AREAS
YOUR EXAMINEES HAVE INCOMPLETE OR UNSTABLE INFORMATION IN THESE CONCEPT AREAS

THOROUGH REVIEW NEEDED IN THESE AREAS

HAVE YOUR WORKSHOP COORDINATOR PREPARE A REVIEW

ITEM	NUMBER	PERCENT	CONCEPT DESCRIPTION/CROSS REFERENCE
17	10.00	0.45	ESTIMATE TEMPERATURE USING DEGREES CELSIUS
			ADDISON WESSLEY 4TH GRADE CHAPTER 6 P 151
16	8.00	0.36	ESTIMATE WEIGHT USING GRAMS AND KILOGRAMS
			ADDISION WESSLEY 4TH GRADE CHAPTER 6 P 150
•	•	•	•
•	•	•	•

INFORMED CONCEPT AREAS
EXAMINEES GENERALLY HAVE REPLICABLE (ACCURATE AND CONFIDENT) INFORMATION IN THESE
CONCEPT AREAS

WORKSHOP COORDINATORS SKIP OVER THESE CONCEPTS OR ADDRESS VERY LIGHTLY

ITEM	NUMBER	PERCENT	CONCEPT DESCRIPTION CROSS REFERENCE
9	22.00	1.00	TO IDENTIFY MULTIPLES OF 1 DIGIT NUMBERS
			ADDISION WESSLEY 4TH GRADE CHAPTER 4 P 94-95
8	21.00	0.95	TO COUNT AMOUNTS OF MONEY
			ADDISON WESSLEY 4TH GRADE CHAPTER 2 P 40-41
•	•	•	•
•	•	•	•

Figure 4
Feedback #2: For Curriculum Fine Tuning
Class Information Needs Profile (CINP)
Generated with the IRT Procedure
(common information needs of a classroom of
elementary school mathematics students)

SINP

MISINFORMED EXAMINATION ITEMS FOR YOUR EXAMINEES
YOUR EXAMINEES HAVE WRONG INFORMATION IN THESE CONCEPT AREAS - MISINFORMED

HAVE WORKSHOP COORDINATOR DEVELOP INSTRUCTIONAL MATERIALS
DEMONSTRATE MISCONCEPTIONS - FOLLOW THIS WITH ACCURATE INFORMATION

ITEM	NUMBER	PERCENT	CONCEPT DESCRIPTION/CROSS REFERENCE
42	182.00	0.46	EVALUATION - MASLOW'S HIERARCHY OF NEEDS
			AC 60-14 P 16/CHOICE AND CHANCE P 184
5	101.00	0.25	SIGMET'S DEFINITION
			AC 00-45C P 4-17/PSG P 18 QUESTION 11
.	.	.	.
.	.	.	.

UNINFORMED TEST ITEMS ON THE EXAMINATION
EXAMINEES GENERALLY LACK INFORMATION IN THESE CONCEPT AREAS

BASIC INSTRUCTION NEEDED

ITEM	NUMBER	PERCENT	CONCEPT DESCRIPTION/CROSS REFERENCE

PART INFORMED CONCEPT AREAS
YOUR EXAMINEES HAVE INCOMPLETE OR UNSTABLE INFORMATION IN THESE CONCEPT AREAS

INSTRUCTIONAL MATERIALS ARE INCOMPLETE OR LACK SUFFICIENT DETAIL

ITEM	NUMBER	PERCENT	CONCEPT DESCRIPTION/CROSS REFERENCE
6	91.00	0.23	MAKING-SHEAR ZONE
			AC 00-06A P 88/PSG P 17 QUESTION 3
43	82.00	0.21	EVALUATION - MASLOW'S HIERARCHY OF NEEDS
			AC 60-14 P 16/MODERN PRACTICE - ADULT EDUCATION 28
.	.	.	.
.	.	.	.

INFORMED CONCEPT AREAS
EXAMINEES GENERALLY HAVE REPLICABLE (ACCURATE AND CONFIDENT) INFORMATION IN THESE CONCEPT AREAS

INSTRUCTIONAL MATERIALS ARE GOOD IN THESE AREAS

ITEM	NUMBER	PERCENT	CONCEPT DESCRIPTION/CROSS REFERENCE
4	385.00	0.97	ORAL QUESTIONING
			INTO PRIV PILOT PTS INTRO PAGE IIII AND IV
3	384.00	0.96	ABILITY TO PERFORM TASK
			PRIV PILOT PTS INTRO PAGE IV
.	.	.	.
.	.	.	.

Figure 5
Feedback #3:
School Information Needs Profile (SINP) -- Resource Allocation
Generated with the IRT Procedure
(common information needs of students assessed by a
government regulatory agency)

student as well as the *common* information needs of all students assessed. Both types of feedback are essential for support in instructional programs, especially at educational sites with large numbers of middle and low achieving students.

4 What Are the Limitations of Present Testing Practices

Policy studies that deal with the *"what"* of presently used testing and assessment practices, tend to underscore the severe limitations of presently used CRT and NRT formats for supporting classroom instruction. The search for alternative testing and assessment formats such as the Information Referenced Testing (IRT) format are meant to address some of these concerns, especially the formative evaluation needs of students, parents, classroom teachers and school administrators.

The following is a list of just a few of the major concerns expressed by educators, with presently used CRT and NRT assessment formats.

- Teachers are not provided with feedback or information in a format needed for instructional decision making and to support instruction.
- At many inner city schools, with large numbers of low and middle achieving students, nearly 1/3 of the students test at or score *below the random chance* or expectation level, indicating a sabotaging strategy or inordinate levels of random guessing.
- The correction for guessing (CFG) formula assumes that all wrong answers on a test are the result of random guesses. This CFG formula thus over corrects and yields a *biased estimates downward* of actual student attainment. Presently only one test, the Scholastic Aptitude Test (SAT) uses CFG formulas for scoring.
- When corrections for guessing are not made, the estimates of student attainment are *biased upwards*. Students actually know less than the score indicates. This inflation of score places them at greater risk of having to build knowledge upon an unstable or unreliable information base in future educational sequences. (See research of Doscher and Bruno 1981, and Wick 1983.)
- The one-dimensional Right or Wrong (RW) test scoring system is in reality a recognition (and non recognition) only type of scoring system and promotes a student test score maximizing strategy that necessitates a guessing strategy in order to increase score.
- There is *no partial credit* given with presently used multiple choice testing and one-dimensional scoring formats. From a student perspective this is an unfair and unnecessary limitation. This limitation also leads to large disparities between the student test score maximizing strategy (maximize score by guessing) and teacher conforming strategy (only answer what you are sure of). The disparity in these strategies is inverse to student attainment levels and leads to erroneous individual assessments of attainment.
- Because of large amounts of misinformation, lack of information and partial information, presently used NRT's and CRT's, scored with one-dimensional R-W test scoring systems, are *insensitive* or yield erroneous

information for instructional decision making when these results are used with low and medium attaining students.

- Misinformation or knowing something wrongly (1/2 + 1/3 = 2/5) and lack of information or knowing nothing cannot be assessed with one-dimensional scoring practices. Thus the important *pedagogical* strategy of "reeducation" for misinformation and "instruction" for lack of information cannot be addressed.
- Current CRT and NRT testing practices are based on the assumption that the classroom *signal of instruction* is perfect and that all imperfections in instruction are associated with the student. In essence, instructional validity or the quality of the instructional program is neither considered, nor is it assessed.

In summary, there are three major concerns regarding the "what" of testing: (1) the conceptual foundations for the test -- normal curve distribution, selection standards, etc. (2) the way students are evaluated against this standard (percent score, percentile, etc.) and (3) the quality of feedback provided to support instructional programs.

5 Research on Alternative Assessment Procedures

A large amount of educational research dealing with two-dimensional and confidence weighted scoring systems (Rippey, 1983, 1986) and "reproducible" confidence weighted scoring systems has been reported in the literature (Anderson 1982, LeClercq 1982, and LeClercq D. and Ph. de Brogniez 1990, Shuford 1966, and Shuford and Brown 1975). Most of this literature is generally positive with regard to both psychometric reliability and validity and basically supports the efficacy of two-dimensional approaches to assessment. For an in depth analysis of the research literature of binary vs. confidence weighted scoring systems see the work of Poizner (1978) in the U. S. A. and LeClercq in Belgium (1980). Software packages to provide two-dimensional assessments have been developed by Shuford, LeClercq and Hunt. All these two-dimensional approaches to assessment are currently being used in a variety of educational contexts.

Extensions of reproducible scoring systems to make them compatible with modern optical scan technologies and powerful micro-computer systems to specifically generate feedback and formative evaluation reports has recently been developed into an Information Referenced Testing (IRT) format (Bruno 1987a, Bruno 1987b, Bruno 1986). The IRT format is presently gaining in acceptance with various government agencies, private corporations, as well as higher education and elementary-secondary education agencies. (See IRT reports depicted in Figure 3, 4 and 5 that are based on some of these applications.)

Applications to urban inner city student instructional audits (Bruno 1988) and direct research comparisons between RW and IRT formats (Bruno 1989b) have also been published. This latter study by Bruno (1989a) explores the extent of the overestimate of knowledge with one-dimensional NRT formats by examining the variations in the quality of information (two-dimensional) for students who have the same exact grade equivalent scores (one-dimensional).

IRT research has also been extended to other educational areas in the research of Barner (1991, 1992) -- math education -- and Baxter (1989, 1990, 1991, 1993) -- at risk inner city student formative evaluation at large inner city elementary schools and Klentschy (1992) -- accelerated learning.

Extensions of the IRT methodology to servicing the needs of school reform movement (Bruno 1989a), student athletes (Bruno 1989c), critical care nursing and technical training (Mathewson 1990) and test retest reliability (Albedi and Bruno 1991) have also appeared in the research literature.

The optical scanning needs and capabilities of the IRT format are depicted in Figure 2. Note that both the simplified version (7 response options used in elementary schools and institutions where student inservice to the IRT procedure is not provided) and the triangle version (13 response options used in most other applications, but some student inservice to the procedure is required) can be used interchangeably by the same IRT computer software package.

The individual education plan (IEP) (see Figure 3) are the "audits" of the student information base. Note from the results of these audits the wide differences in quality of information possessed by these two critical care nurses working at the same hospital. The audit concept is used to provide a plan of action to keep the student (or employee) "on line" with regard to currently needed reliable information. In this particular illustration the audits suggest that some critical care nurses are dangerously outdated with regard to their levels of reliable (accurate and confident) information (see Mathewson 1990). Note that instructional cross references are also provided on the IEP in order to stimulate student information seeking behaviors. This IRT report called the Individual Education Plan or IEP essentialy articulates the instructional program to students (and parents) and is used solely for purposes of formative evaluation.

Figure 4 illustrates the common information needs profile (CINP) across all students in a particular classroom or instructional unit (a classroom setting). Note that this feedback ranks the most to least misinformed concepts, the most to least uninformed concepts, etc. This particular illustration (Figure 4) was drawn from an experiment at a local elementary school in the area of pre-algebra. Note how teachers can use this IRT report information to prepare class reviews and to fine tune the school curricula to the exact information needs of a particular classroom of students. Instructional cross references are also provided in the CINP.

Figure 5 depicts how the IRT procedure is used to identify common information needs across all students and all instructional units. The School (or System) Information Needs Profile (SINP) is used to examine how instructional materials and teaching effectiveness (the signal) is being received by all students. This IRT report is used by educational administrative personnel to identify inordinate amounts of misinformation (suggesting the rewriting of instructional materials), lack of information (suggesting the developing of supplementary instructional materials), and partial information (suggesting the adding of greater clarity and comprehensiveness to instructional materials) in an instructional program.

Other IRT generated reports include a confidence contour analysis (CCA) or a two-dimensional item analysis. This report is used to identify the frequency of use associated with each choice and the relative confidence in that choice. This IRT report can be used by educational evaluators for purposes of redesigning the assessment instrument.

The distribution of information report (DI) examines the quality of information for each concept across instructional units. This report can be used by school administrative personnel to identify weak instructional units. Staff development and inservice programs can then be designed to address these needs.

These are several IRT Summative Evaluation reports that are generated by the software package. These reports contain a two-dimensional IRT score (percent of perfect information); average confidence in correct information (based on the IRT score); student self-appraisal of their information (what they expect to receive as a score)and a measure of student realism (percent correct when 100% sure of an answer).

All of these IRT reports, directed at various elements of the instructional leadership TRIAD -- students, teachers, administrators, can be used to provide formative evaluation feedback to support instructional programs. Areas of curriculum design, curriculum fine tuning, curriculum articulation, individual education plans, resource allocation to instruction, design of staff development and inservice programs, redesign of the assessment instrument, and reports for purposes of summative evaluation (selection and certification) are all efficiently and effectively addressed with information provided by the IRT procedure.

6 Conclusion

In conclusion, educators will always have problems assessing "true" student academic attainment in the classroom and providing detailed and accurate feedback to support the instructional process. Biochemistry has not advanced to the state where a student will produce an enzyme when he or she has learned, say algebra or French or science. If this were ever the case teachers, by means of a simple blood test, could then clearly and unambiguously determine if actual learning had taken place. Until these advances in biochemistry come to fruition, however, all educational assessments will have to be "passive" or "one step removed" types of assessments. In essence, student responses to stimuli have to be evaluated as a proxy for measuring learning.

What seems to be needed in education is a total re-examination of the basic philosophy of the "why" we test. The philosophy presented here is that most classroom testing should be used to promote learning and support instruction, especially for low and medium attaining students. The way medical doctors use testing to promote the health of a patient, teachers should use testing to promote the learning of the child. The way medical testing provides detailed feedback to the patient and the hospital and is used to monitor health and suggest strategies for improving health, educational testing should monitor learning and suggest appropriate instructional interventions to enhance student learning.

The *Information Referenced Testing* concept, briefly presented here, is directed specifically at addressing the formative evaluation needs of an *individual* student as well as the *common* information needs of a classroom of students. As other multiple choice "objective" test scoring formats (answer until correct, ordinal scoring, etc.) and subjective test scoring formats (portfolios, performance testing, etc.) become available, they should all be evaluated against the economic, legal as well as the educational standards for formative (feedback) as well as summative evaluation. The important relationship between assessment and student learning has to be addressed with any assessment format that is designed to support instruction. Information Referenced Testing is a significant step in filling the important *formative* evaluation niche in our assessment practices. IRT might provide the type of information "sensitive" assessment instrumentation that is needed to service the challenging demands of the school reform movement in America and to provide the type of feedback needed to support instructional programs in educational organizations.

References

Albedi, Jamal and James E. Bruno (1989) *Test-retest reliability of computer based MCW-APM test scoring methods.* Journal of Computer Based Instruction 16(1), 29-35.

Anderson, Richard (1982) *Computer-based confidence testing: Alternatives to conventional, computer-based multiple-choice testing.* Journal of Computer-Based Instruction 9(1), 1-9.

Barner, Robert (1992) *Formative evaluation and student attainment at inner-city schools.* Unpublished Doctoral Dissertation, University of California, Los Angeles, CA.

Barner, Robert and James E. Bruno (Submitted for review, 1990) *Formative evaluation and mathematics attainment of inner city students.* Urban Review.

Baxter, James B. (1993) *Formative evaluation and the self-assessment process to support instruction at inner city schools.* Unpublished doctoral dissertation, University of California, Los Angeles, CA.

Baxter, James B. (1991) *Information referenced testing for formative evaluation and self-assessment to support the elaboration and advancement of student thinking and learning.* Paper presented at the Eighth International Conference on Technology and Education, Toronto, Canada.

Baxter, James B. (1990) *Information referenced testing for formative evaluation.* Paper presented at the Seventh International Conference on Technology and Education, Brussels, Belgium.

Baxter, James B. and James E. Bruno (1989) *Computer assisted learning with information referenced testing.* Proceedings, International Symposium on Computer Assisted Learning, University of Surrey, England.

Bruno, James E. (1986) *Assessing the knowledge base of students: An information theoretic approach to testing.* Measurement and Evaluation in Counseling and Development 19(3), 116-130.

Bruno, James E. (1987a) *Admissible probability measurement in instructional management.* Journal of Computer Based Instruction 14(1), 23-30.

Bruno, James E. (1987b) *Using computers for instructional delivery and diagnosis of student learning in elementary schools.* Journal of Computers in the Schools 4(2), 117-134.

Bruno, James E. (1988) *The instructional audit in urban school settings: A descriptive policy analysis of instructional delivery using signal-receptor assessment theory and information referenced testing.* The Urban Review 20(2), 95-107.

Bruno, James E. (1989a) *School reform: A thermostat instead of a thermometer.* In: S. Cohen and L Solmon (eds.) From the Campus. New York: Praeger.

Bruno, James E. (1989b) *Monitoring the academic progress of low-achieving students: An analysis of right-wrong (R-W) versus information referenced (MCW-APM) formative and summative procedures.* Journal of Research and Development in Education 22(4), 51-61.

Bruno, James E. (1989c) *Developing a professional standard of care for academic advisors of student athletes: Use of information based formative evaluation in academic support.* The Academic Athletic Journal 17-35.

Doscher, Lynn and Bruno, James E. (1981) *Simulation of Inner City Standardized Testing Behavior: Implications for Instructional Evaluation.* American Education Research Journal 18(4), 475-489.

Hunt, Darwin P. (1977) *The human self-assessment process. Study II: The effects of the number of self-assessment categories on acquisition.* Interim Report from U. S. Army Research Institute for the Behavioral and Social Sciences Grant #DAHC19-76-G-002, New Mexico State University, Las Cruces, NM.

Hunt, Darwin P. (1982) *Effects of human self-assessment responding on learning.* Journal of Applied Psychology 67(1), 75-82.

Hunt, Darwin P. and Michell R. Sams. (1989). *Human self-assessment process theory: An eight factor model of human performance and learning.* In: G. L. Ljunggien, S. Darnic (eds.) Psychophysics in Action, pp. 41-53.

Klentschy, Michael L. (1992a) *Designing instructional support and decision making systems to service accelerated learning environments in large urban elementary schools.* Unpublished doctoral dissertation, University of California, Los Angeles, CA.

Klentschy, Michael L. (1992b) *Supporting accelerated learning with information referenced testing.* 9th International Conference on Technology and Education, Paris, France, March 1992.

Leclercq, D. (1982) *Confidence marking, its use in testing.* In B. Choppin and N. Postlethwaite (eds.) Evaluation in Education, An International Review Series 6(2), 161-287.

Leclercq, D. and Ph. de Brogniez (1990) *A fresh look on confidence marking.* Paper presented at the Seventh International Conference on Technology and Education, Brussels, Belgium.

Mathewson, Sheila (1990) *Designing human resource and staff development programs in information dependent organizations: An application of periodic information audits or critical care nursing personnel.* Unpublished Doctoral Dissertation, University of California, Los Angeles, CA.

Poizner, Sharon, W. Alan Nicewander and Charles Gettys (1978) *Alternative responses and scoring methods for multiple-choice items: An empirical study of probabilistic and ordinal response modes.* Applied Psychological Measurement 2(1), 83-6.

Rippey, Robert, and Anthony Voytovich (1983) *Linking knowledge, realism and diagnostic reasoning by computer-assisted confidence testing.* Journal of Computer-Based Instruction 9(3), 88-97.

Rippey, Robert (1986) *A computer program for administering and scoring confidence tests.* Behavior Research Methods, Instruments and Computers 18(1), 59-60.

Shuford, Emir, A. Albert, and H. E. Massengill (1966) *Admissible probability measurement procedures.* Psychometrika 31(2), 125-145.

Shuford, Emir and Thomas Brown (1975) *Elicitation of personal probabilities and their assessment.* Instructional Science 4(2), 137-188.

Wick, J. W. (1983) *Research Note.* American Education Research Journal 20:461-468.

The TASTE Approach:
General Implicit Solutions in Multiple Choice Questions (MCQs), Open Books Exams and Interactive Testing

D. Leclercq, E. Boxus, P. de Brogniez, H. Wuidar, F. Lambert

Université de Liège, Service de Technologie de l'Education (STE), Batiment 32, Sart Tilman, 4000 Liège 1, Belgium, Tel. 32-41-56 20 72, Fax 32-41-56 29 44, e-mail U017801 at BLIULG11

Abstract: The efficiency of school to efficiently train in higher order cognitive skills is sometimes questioned. Principles and techniques (such as open books exams, implicit questioning, self assessment) have been developed to enhance cognitive vigilance, data processing skills and metacognition. The details of those techniques are provided as well as some results about question characteristics and student opinions. Further developments are considered and perspectivized with ambitious training and assessment goals.

TASTE is an anonym that means "Towards an Adult System of Training and Evaluation". Its rationale and components will be described hereafter.

1 General Implicit Solutions in Multiple Choice Questions (MCQs)

1.1 The recall/recognize issue

It has been frequently advocated that MCQs do not assess the same type of ability as open questions or essays do. In particular, a student can *recognize* among the printed solution of a MCQ the correct answer whereas he/she would not be able to *recall* it.

It has been suggested (Wood, 1977) to add the option "none of those solutions" as the last solution. For the MCQs where this option is the correct one, the student cannot "recognize" but is forced to recall. We have adopted this principle, as well as the option "all of them", referring to Noizet and Fabre (1975) who present the possible solutions of a multiple choice question in a "simplex" structure.

1.2 A taxonomic challenge

It is well known that MCQs insidiously lead teachers to ask questions that deal only with *details*. The reason for this is that on details everybody agree and the correct answer will not be disputed.

Wood (1977) has suggested to use MCQs where the student could answer "I *lack data* to be able to answer" or "there is an *absurdity* in the stem that makes the whole question *nonsense.*" These types of question force the student to consider the relevance of the data or of the question themselves, i.e. to *understand* the problem and *analyse* the way it is stated instead of limiting him/herself at plain *knowledge* and *application* levels (in the terms of Bloom's taxonomy of cognitive objectives). Whe have adopted this principle, making it systematic, i.e. applying it in all our questions.

1.3 The hidden curriculum

The hidden (or latent) curriculum can be defined as "what nobody teaches but everybody learns". It is true that, by law of effect (operant conditioning), feedbacks not only reinforce the selection of the correct answer but also the responding behavior.

In this way, through explicit questioning, students "learn" a series of hidden principles such as :
   - You should answer when questioned (What about questions that do not deserve an answer or that should not be asked ?);
   - You should wait to be questioned to answer (This lowers the tendency to raise questions spontaneously, a very healthy behavior in cognitive functioning).

Through "limpid" and simple questions, students learn that :
   - When the authority (here the teacher) asks a question, it is always a relevant one and there always exists an answer, more specifically *one* answer and *only one* (Whereas we all are frequently presented wrongly formulated questions or questions for which there exists no available answer or others for which several answers can fit).

This hidden curriculum is totally in opposition with real life that requests from persons to "detect problems", to assess likelihood of reasoning, data necessity, data sufficiency, etc.

The gap is so great that only a *systematic and massive counter curriculum approach* can counterbalance the hidden curriculum's negative effects. *Implicit questioning* is an endeavour in this direction.

1.4 Implicit questioning

The so-called Dr. Fox's experiment is a game where a lecturer presents stupidities, nonsense reasoning, incorrect data ... up to the moment a person from the audience stops him/her, questioning the relevance of the content. Often, lecturers can speak for minutes and minutes before being interrupted: we have not been trained in *detecting* awkward reasoning, ill stated problems, etc.

Actually, the majority of the audience is *able* to decide it but *only if they are asked.* Therefore the training must focus of the detection side; that is the reason why we developed implicit questioning, i.e. *the teacher creates a situation* in which he hopes (he expects) the student will react *but without presenting it as a question.*

In our MCQ instructions, students are informed that they will have to consider two types of *solutions* :
   - the *printed* ones;

- additional ones that have been described at the start of the test (and will remain under the student's eyes during the whole testing) but *that will NOT be repeated in each question* : the implicit solutions.

1.5 General solutions

We have limited the implicit solutions to four types that can be general, i.e. stand for each of the MCQs appearing in a school test. These types include:

1. REJECT (R or 6) : None of these proposed solutions is correct.
2. TOTAL (T or 7) : Each of these solutions is correct.
3. MISSING (M or 8) : The stem of the MCQ lacks a piece of information that is necessary to decide which solution is correct.
4. ABSURDITY (A or 9) : The stem of the MCQ contains an illogical statement, i.e., an error, an absurdity that makes the whole question irrelevant.

As a general rule, the 9 answer (A) has priority over all the others : an absurdity has to be detected and pointed out *instead of (and before)* trying to answer the question. It is not sufficient to answer 6 (Reject : None) since, when the stem countains an absurdity, all solutions are automatically incorrect answers.

In the same way, the 8 solution (Lack of data) has priority over the 7 solution (Total) since, when data is lacking, all solutions are *potentially* correct.

Here are a few examples illustrating what has to be answered when those General Implicit Solutions are considered :

Question		Correct Answer
The capital of France is	1. Lille 2. Lyon 3. Paris	3
The capital of Italy is	1. Berlin 2. Praga 3. Tokyo	6
U.K. contains	1. England 2. Northern Ireland 3. Wales	7
How old was Lincoln ?	1. 20 years old 2. 30 years old 3. 40 years old	8
In which year did Napoleon and Hitler meet?	1. 1850 2. 1915 3. 1945	9

1.6 Advantages of General Implicit Solutions (GISs)

The Reject (6 or None) solution forces the student to answer mentally *before* having a look to the suggested solutions, *then* to check among them whether his/her own solution is presented. This transforms the usual "recognition" performance into a "recall" one.

The Total (7 or all) solution trains the student to consider that sometimes there exist several correct solutions to a given problem.

The Missing (or lack) of data (8) solution enables the teacher to assess Bloom's levels of cognitive processes such as analysis and intelligent application, since being able to apply a principle implies to be able to detect when it is impossible to apply it and why.

Last, the Absurdity (9) solution is a good way to assess "deep understanding", namely in detecting contradictions.

Usually, all those GISs are not introduced all together in the instructions for untrained students. A strategy of progressivity is recommended.

2 Confidence marking

It has been shown elsewhere (Leclercq, 1983) that faced to inappropriate scales of tariffs (point awards), students discover that some strategies consisting to bias their intimate estimation of confidence pay more than telling the truth.

In order to reinforce students to tell the truth, to make it the optimal behaviour, tariffs must be computed according to decision theory (Von Neumann & Morgenstern, 1947).

It must be noted that very few researchers have followed these methodological conditions, which has resulted in inconsistent data and finally ruined a fruitful approach, which was almost abandoned during the late 1970s. Fortunately, a few scholars have continued to promote strict methodological requirements, what De Finetti (1956) calls "Methods for discriminating levels of partial knowledge ...", what Shuford et al. (1966) call "Admissible probability measurement procedures", and what Van Naerssen (1962) calls "A scale for the measurement of subjective probabilities". To these pioneers must be added Darwin Hunt and James Bruno, the systems of whom are described in Leclercq and Bruno (1993).

Actually, five fundamental methodological conditions should be fulfilled to use confidence marking in a valid way.

2.1 The instructions must offer a metric scale

Most researches have used vague (ordinal) instructions such as "Tell me whether you are *strongly* sure, *fairly* sure, *weakly* sure about your answer". Ordinal processing of data are spurious since we have no guarantee that even within a single person those "yardsticks" keep the same meaning from test to test, from item to item. Comparisons between different persons are, of course, to be excluded for the same reasons.

Here are instructions specifiying codes to designate defined portions of the probability scale. Experiments show that (like in differential thresholds in psychophysics), sensitivity is not the same at the extreme portion of the scale than in the middle (Leclercq, 1983, 1992).

2.2 Tariffs must be computed according to decision theory

It is not sufficient to score with higher points a correct answer with a high confidence degree than a correct answer with a low confidence degree. All points (positive and negative ones) must be computed so that the learner is interested in expressing his doubt with realism and without bias i.e. in telling the truth or admitting to their actual knowledge. The series of tariffs hereafter is an example of such a scale, that insures local optimality of each confidence degree (see Figure 1).

For the previous scales, the tariffs we adopted were as follows:

Confid. code:	0		1		2		3		4		5	
	0	25		50		70		85		95		100
Tariffs:												
correct	13		16		17		18		19		20	
incorrect	4		3		2		0		–6		–20	

A series of other "acceptable" scales and tariffs have been described elsewhere (Leclercq, 1983, p. 210).

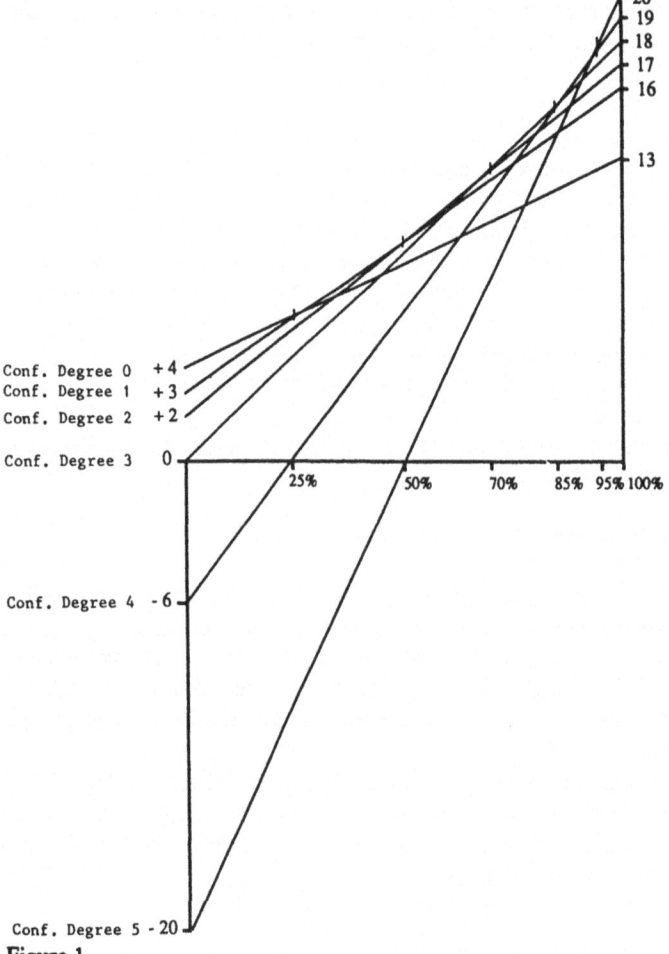

Figure 1

This graphic is the result of an operational research. It has to be "read" as follows. For each confidence degree, the "expected scores" (a theoretical concept) can be computed for any value of subjective probability, by the formula:

expected score (where using degree i with subjective probability being p)
= $(p.TC_i) + (q.TI_i)$

where TC_i is the Tariff (amount of points for i) in case of Correct answer
TI_i is the Tariff (amount of points for i) in case of Incorrect answer
(for instance, TC is +20 and TI is –20 for confidence degree 5).
p is the (subjective) probability of the learner (i.e. his/her confidence in his/her answer being correct).
q is the subjective probability of the learner's answer being incorrect.
Actually, q is $1 - p$ (if p is 0.8 then q is 0.2).

The expected score computed for all the values of p constitute a straight line. It can be drawn without computation since, for instance for confidence degree 5, the expected score is - 20 when p is 0 (the min.for p) and + 20 when p is 100 (the max for p). Joining those two extremes values creates the line of expected scores for each confidence degree.

In figure 1, the e.s. lines have been drawn for each of the six confidence degrees. It can be seen that there is only a definite sector where each e.s. line is optimal (i.e. overtops all the other e.s. lines), and this sector corresponds to what has been announced in the instructions, i.e. the limits (in probability terms) of confidence zones. This shows "optically" which confidence degree is *optimal* (i.e. gives the highest exptected score) for which probability.

Such a diagram has to be computed for each kind of segmentation of the probability scale. It can be drawn by trial and errors, but it may take time! A computer program has been written to insure this calculation (Leclercq, 1983, p. 200).

2.3 We must distinguish measurements from payments

Numerous researches, most of which were published in the Journal of Educational Measurement, raise the question : "Are new (total) test scores (computed with new scales of tariffs taking confidence degrees into account) more valid and more reliable than classical ones (number of correct answers) ?"

Results from these experimental results are confusing. Half of the studies show an increase in validity and a decrease in reliability whereas other studies find the contrary ... without being able to explain why.

Actually, the problem itself is incorrectly stated since the new total score is not a measure, but the combination of two different measures :
- the measure of ability (number of correct answers);
- the measure of realism (quality of self assessment).

The new score can be more valid (i.e. reflect more accurately the learner's competency) only if the person is realistic!

2.4 Explicit feedback about realism must be provided

The CERT computer program computes several mathematical indices and outputs a graphic representation of realism. The following figures (2,3 and 4) present 3 examples of students' "realism" when using instructions hereover.

The student on the left (2) overestimates himself, the student of the right (4) underestimates and the student of the middle (3) is very well calibrated.

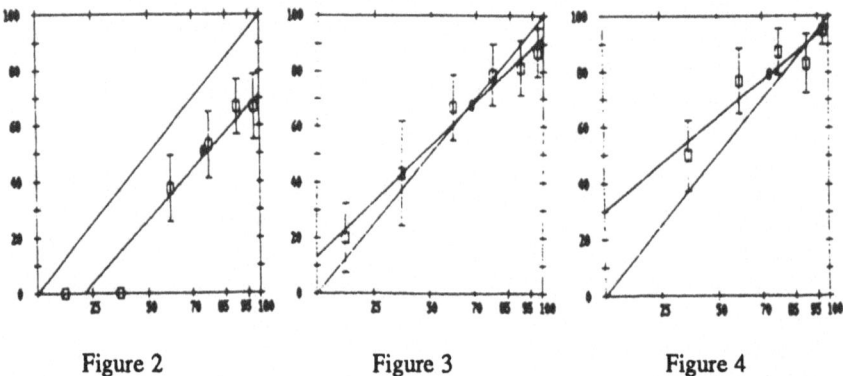

| Figure 2 | Figure 3 | Figure 4 |

Each little "square" represents a rate of success for the given portion of the axis, i.e. for the five confidence zone. Each square is accompanied by the interval of confidence (standard error of measurement) of the percentage it represents (formula = pq/N).

If the error range (drawn over and under the square) overlaps the diagonal line (that represents perfect realism), we cannot conclude that the student is unrealistic.

2.5 Students must be trained

When they use confidence degrees for the first time, students usually adopt non optimal strategies (well known by decision making specialists, such as Lindley, Raiffa, Savage, Luce, Tversky, Kahneman, who have even given names to those strategies).

- choose the degree with the highest score (in case of success), i.e. an optimistic strategy (actually called *maximax*));.
- choose the confidence degree that insures the lowest "lost" (in case of failure), i.e. pessimistic strategy (actually called *maximin*);
- a mixture of the two (actually called Hurwicz's principle);
- always the same degree according to a fixed probability (for instance always degree 2, according to 60 % of chances) (actually called Atkinson's principle);
- give the probability corresponding to random guessing, i.e. the principle of total ignorance (actually called Laplace's principle);
- give the confidence degree that provides a given Variance (or difference) between the two tariffs, i.e. in case of success and in case of failure (actually called Coombs' principle);
- etc.

None of these strategies is optimal.

Neither verbal nor graphic nor equational explanations can convince the students that telling the truth is the best strategy in terms of maximizing their points. *They have to experience it.*

Only after the first trial, and above all the first *feedback,* do they admit a series of evidences: their strategy was bad, and their neighbours' strategies were even worse.

2.6 Adapt to situations

Sougne et al. (1990) suggested an original mode of answering to a multiple choice question :

- for each alternative (here 5), answer by T (true) or false (F);
- then, give *ONE confidence degree of this set of (5) answers* using the following probability scale (recommended by Leclercq, 1988).

Figure 5 illustrates this.

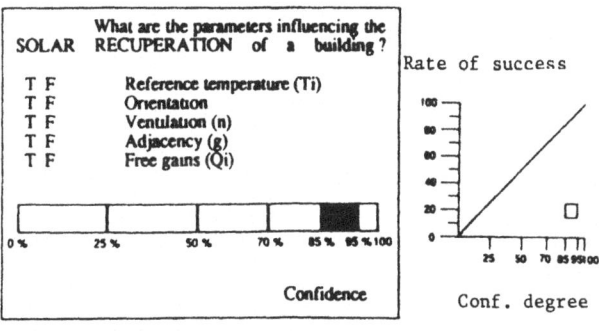

| Figure 5 | Figure 6 |

The advantage of such an approach is that an index (and a graphic of realism are available for each series of questions. Figure 6 illustrates the graphic for a person who answered 1 question correctly out of 5 (20 % correct) with a confidence degree of 90 %.

Figure 7 illustrates more classical alternative instructions and answer modes:

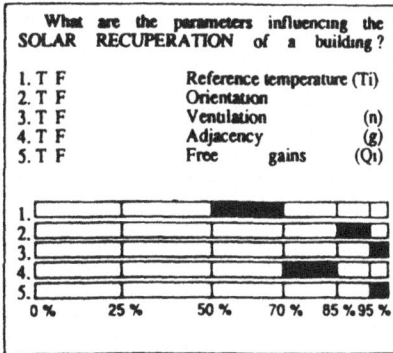

Figure 7

Of course only 20 groups of questions of this type provide 100 (T/F) answers and confidence degrees, enabling fruitful statistics.

The five principles described in this section have been followed in the experience that will be described:
  - metric scale, asymetric to fit human sensitivity (acuity) and reliability
  - tariffs computed according decision theory principles to reinforce the honest expression of probability estimates (i.e. telling the truth)
  - distinguishing measurements (of competency and realism) from the payment of what the learner observes (actually a weighted combination of two measures).
  - providing explicit feedback on realism with the help of a graphical representation (called the calibration graphic).
  - familiarizing students with the system.
  - adapt to situations: in some occasions, the learner will have to give one confidence degree for one single answer, in other occasions he/she will have to distribute his/her 100 % probabilities among alternatives, in other occasions, he/she will have to give only one probability value for a series of related answers.

The use of confidence degrees is not a necessary characteristic in the TASTE approach, but it has been incorporated in it from the beginning.

3 Open-book exams (the TASTE approach)

Open-book work is the most common situation in adult work, as well as in adult learning. Often, we discover new things to understand and memorize from written text. We read them, we try them (on the computer, the engine or the objects dealing with the content) and, *only if we do not understand,* do we ask others.

Since training can more and more rely on (well written and well illustrated) textbooks, videos and computer programs, it is sensible to apply within training the principles of adult learning conditions, in a formula called TASTE : "Towards an Adult System of Training and Evaluation".

The various steps of TASTE are as follows.

Step 1. *Before the course/exam day*

Learners (at home) read books, watch videos, etc.

Step 2. *On the course/exam day*

Learners ask the trainer questions about contents which were not understood. Those questions are mostly of two types.

The first ones adress prerequisite notions they are lacking. That is the bad news for the trainer (he/she is forced to re-teach unmastered contents), but for those learners it is of paramount importance that theses requests be met.

The second type of questions focus to the "top" of the content: ambiguities within one document, contradictions among various documents, connexions with quite different contents, future development, etc. This is the good news for the trainer because those interactions are rich, even for the teacher! He/she has sometimes the feeling of discussing with peers (colleagues), or at least with informed persons, what is scarcely the case in the classical lecturing situation.

Step 3. *The exam*

The trainer distributes the questions; learners answer on two sheets: one they give back to the trainer (who will introduce the data into the computer) and one (identical) they keep for themselves. For each question, students are invited to provide three types of responses in a preprinted sheet : their answer, a confidence degree (see Leclercq, 1992) and justifying comments.

Q	A	C	Justification and comments
1			
2			
3			
4			
5			
Etc			

Q = Question   A = Answer   C = Confidence degree

Figure 8: Answer sheet for a student

Step 4. *Direct feedback*

The trainer/teacher communicates the correct answers and each student is free to oppose it, i.e. to argue in favour of other answers he/she considers should be accepted as correct. If the teacher is convinced of the correctness of the reasoning, this one is given more importance than the chosen coded answer, so that the teacher changes the given answer to the correct one ... provided the argument has been written in the justifying comments area of his/her answer sheet.

This step has immediate effects on the quality of questions (the teacher benefiting from remarks about each question), as well as the unambiguity of the reference document (it is sometimes questioned as well).

Some questions may reveal themselves to be so ambiguous that the teacher may immediately decide to suppress them so that the students' scores will not be affected by those wrongly phrased questions.

4 Interactive testing (The CHECK package)

CHECK is a series of interrelated softwares (mainly written in CLIPPER by H. Wuidar and partly in PASCAL by Ph. De Brogniez) that offer some evaluation facilities to the teachers and to the students.

4.1 CHECK IN

This software enables the teacher/trainer to add new questions (or modify existing questions) in one of his/her item banks.

Each question has the following structure :
- a code number;
- the phrasing of the objective (that will be printed on the students's report sheet);
- the phrasing of the question itself (with the suggested solutions);
- the instruction code (since several instructions concerning the type of question, the mode of answer and the tariffs are available);
- the correct answer code;

- a feedback (2 lines max.) that will be displayed on the screen just after the student has confirmed his/her answer and is told the correct one;
- a content code (optional);
- a mental process code (optional).

4.2 CHECK FROM

This software enables the teacher to select questions from the item bank to constitute (automatically) parallel versions either of paper-pencil tests or of an interactive test.

4.3 CHECK UP

This software enables each student to choose a given domain (within one of the various item banks) and to train (or practice) in answering a series of questions, being noted and receiving (immediate or delayed) feedback.

The successive steps of the test are as follows.
1° The student is told how many items are in the selected test.
2° The student is informed that
- all the questions are presented twice to him/her, but in a fixed order;
- during the first presentation, the student has to provide an answer and a confidence degree that is NOT scored;
- during the second presentation, the student has the possibility to change his/her previous answer and confidence degree, but they are now definitive and are not scored;
- the instructions, especially the codes for Reject (6), All (7), Lack of data (8) and Absurdity (9), are permanently displaied on the top of the screen;
- time spent is permanently displaied on the (top of the)screen;
- the clock is stopped between two questions (during this period, the students are invited to write their justification on a piece of paper).
3° The testing starts, item by item, for a first presentation where
- the question is displayed (and chrono starts);
- the learner answers (response + confidence degree), then is asked to revise or confirm.
4° The second presentation is the same except that, after each confirmation of the answer and the confidence degree, chrono is stopped and the learner is asked whether the correct answer and his/her scores should be communicated to him/her immediately (step 5) or at the very end of the test (step 7).
5° The correct answer, the feedback and the total score are displayed (if wanted). The student fills (handwriting) the justification area on his/her answer sheet.
6° Next question is displayed and chrono starts again.
7° When all questions have been answered, if wanted, all the feedback not yet communicated, are displayed one after each other.
8° At the end of the testing, a print out is produced with the help of CHECK OUT (see the following example). Nevertheless, the student is informed that the teacher will read his/her handwritten justifications and could change (favorably since justifications for correct answers will even not be read at all)

the report by introducing the correct answer for some incorrect responses and by producing a new report.

4.4 CHECK OUT

The software enables to print various types of reports. Here is an abbreviated version of such a report.

		FIRST RUN				SECOND RUN			
Quest.	Ans.	Conf	0/1	Time	Ans.	Conf	0/1	Time	Objective
1	3	5	0	136	3	2	0	127	Understand redacti
2									Understand what a
3									Detect redaction fl
4									Understand technic
5									Apply COOMBS, MI
6									Understand the imp
7									Apply SLAKTER's ap

Figure 9: Extract from a feedback sheet provided to a student after a CHECK sesssion

As can be seen, the student is not given the items themselves, to prevent students from learning the questions instead of the course.

5 Results

F. Lambert (1992), made the following observations on tests that were used for different chapters of two of Leclercq's courses (course 155 with 20 students, course 58 with 120 students). Suppressed questions are those that the teacher decided, either on the basis of the students answers or comments, or on the basis of item analysis, to suppress after the students had answered.

5.1 Mean time to answer and status of the questions

Type of questions	C155 Test 1 20 students 12 questions	C155 Test 2 20 students 15 questions	C155 Test 3 20 students 12 questions
Correctly answered	162	245	205
Incorrectly answered	215	277	276
Suppressed	262	295	307

Figure 10: Mean time in seconds for answering

Repeatedly, in this course (on evaluation) correctly answered questions need less time than incorrectly answered ones and far less than suppressed ones. So testing time is reduced by the students' abilities (less incorrect answers) and by the questions quality (less suppressed questions).

5.2 Time to answer and type of question

The following observations are averaged from test 1 (122 students of second year of graduate studies in psychology answering 12 questions) and test 2 (120 out the same students answering 12 other questions) of course 58 (educational technology).

Out of those 2904 answers, only 2134 (73%) have been considered because they have not been modified after the teacher has read the students' justifications.

Average data for the 4 types of general implicit solutions (GISs) are as follows :

	Number of answers	Time in seconds	Rate of success
Usual	708	269	68 %
Reject (6)	399	297	59 %
Total (7)	406	226	66 %
Missing (8)	245	229	57 %
Absurdity (9)	376	179	74 %

Figure 11: Statistics concerning 5 types of solutions

It appears that the *Absurdity* solution is the quickest and the most correctly answered. *Lack* of data has the lowest rate of success and *Reject* takes the longest time to answer.

5.3 Discrimination indices

The previous results deserve to be compared to the observations made in may 1987 on 140 students (fourth year of graduate studies, from various orientations : sciences, linguistic, social sciences, economics, etc.) having answered 82 MCQs with 4 kinds of SGIs :

Correlation	Number of quest.	Aver. success	Aver. point biserial
Usual	40	66 %	0,29
Reject	9	48 %	0,38
Total	11	55 %	0,31
Missing	10	28 %	0,29
Absurdity	10	66 %	0,42

Figure 12: Average characteristic of 5 types of items

Here again, the most difficult appeared to be the *"lack of data"* type of question and the absurdity the most frequently correct (with the usual type of questions).

About the same order in average difficulty is observed in the two experiments.

ABSURDITY and REJECT solutions have the best point biserial correlation indices.

It is worth nothing that *the average discrimination indices of GISs are higher* than the usual questions ones!

5.4 With and without book

As a part of the 1988 experiment, 22 questions had to be answered twice : the first time without the book and the second time with the book. There were four questions of each of the five types of questions, plus two fictitious questions. Those fictitious questions concern inexisting contents, according to Slakter, Koehler and Hampton 's idea (1970). We considered, as correct answers for those questions (non announced in the instructions) three answers : "Omit", "Lack of data" and "Absurdity".

Here are the results (on 140 students) :

Categories	Number of questions	Main facilities		
		without the book	with the book	GAIN
Fictitious	2	47 %	62 %	+15 %
Usual	4	50 %	54 %	+ 4 %
Reject	4	40 %	48 %	+ 8 %
Total	4	55 %	60 %	+ 5 %
Missing	4	33 %	35 %	+ 2 %
Absurdity	4	58 %	65 %	+ 7 %

Figure 13: Average characteristics of 6 types of items

As foreseen, the *greatest gain* happens for the *fictitious* questions (the book had an index of keywords and an other index of authors names).

After that, *REJECT and ABSURDITY benefited the most from the book*, and LACK benefited the least.

Gains are deceptive in average. This can be partly explained by the fact that maybe students have not be given enough time (browsing into a book requests a great amount of time) and by the fact that most of those students were not trained to this kind of answering (to those instructions).

6 The students' opinion

F. Lambert (1992) asked to the 140 students (second year of graduate studies in psychology and education) of course 58 (educational technology) to answer an opinion questionnaire. Here are the results to various questions:

6.1 The FAIREST type of questions (N = 119)

Students had to grade each question on a 10 degrees scale about which system is the fairest. Average results are as follows:

MCQ with GIS and open books and justification comments	6,53
Written exams with long answers	5,22
Oral exams	5,15
Written exam with short answer	5,22
Classical MCQs	4,31

Figure 14: Students' opinions about several types of questionning methods

F. Lambert has been very cautious to avoid the "desirability phenomenon" (written commitment not to reveal to the professor who has answered what). Nevertheless, this phenomenon is still likely to have occurred. The following results to other questions could nevertheless help understand the causes for the preferences.

6.2 Appreciations of justifications (N=124)

Students were asked:
   a) whether they *strongly dislike* (1) or *strongly like* (7) (on a 7 positions scale) the principle of their writing justifications, these ones been read by the teacher only in case of incorrect answer and constituting a second chance for being credited with a correct answer;
   b) whether (7) or not (1) the possibility of winning points by written justification compensates time it needs to write them down.
   Results are as follows:

a			b	
I strongly dislike	1	4.0	Does not compensate	6.6
	2	7.3		9.0
	3	4.0		7.4
	4	8.1		13.1
	5	12.9		13.9
I strongly like	6	21.8		21.3
	7	41.9	Compensates	28.7
		100.0		100.0

Figure 15: Students' opinions about the justification principle

Subhypothesis like "the students who actually use *and benefit* from this piece of instruction are more favorable than others" have not been tested yet (whereas data would permit that).

6.3 Preferences for oral or written exams (N=121)

Those two possibilities were the opposite "poles" of a seven degrees scales.

oral	1	9,9 %
	2	14,9 %
	3	13,2 %
	4	21,5 %
	5	20,7 %
	6	14,0 %
written	7	5,8 %

Opinions vary broadly, no consensus is possible. This is very important for examiners to know that *probably there does not exist a single instruction that will fit all students !*

Therefore, one should consider instructions that can be adapted to the learner's preferences on his/her own request.

6.4 Preferences for open (long) answer questions or MCQs (N=121)

Open long answers	1	3,3 %
	2	13,2 %
	3	5,0 %
	4	11,6 %
	5	18,2 %
	6	27,3 %
MCQs	7	21,5 %

Whereas there is a tendency of a greater appeal of MCQs, preferences are spread along the whole continuum.

6.5 Preferences for Open (short answer) questions or MCQs (N=120).

Open short answers	1	5,8 %
	2	11,7 %
	3	12,5 %
	4	22,5 %
	5	13,3 %
	6	22,5 %
MCQs	7	11,7 %

Opinions are largely spread on the continuum and the advantage for MCQs is lower than it was for long answers to open questions.

6.6 Among MCQs, preferences for GISs or classical (N=122)

Classical MCQs	1	45,9 %
	2	18,9 %
	3	11,5 %
	4	5,7 %
	5	3,3 %
	6	7,4 %
GIS MCQs	7	7,4 %

Preferences are largely in favor of classical MCQs. Note that those students have never been shown statistical data about rates of success of the five types of questions.

6.7 How far is each of the SGIs appreciated?

	None reject	All of them	Lack of data	Absurdity
	6	7	8	9
1. strongly dislike	12,1	8,4	44,5	34,2
2.	6,5	4,2	18,5	13,3
3.	14,5	14,3	11,8	10,8
4.	31,5	31,9	10,9	13,3
5.	25,8	26,1	5,0	15,8
6.	7,3	10,1	6,7	5,0
7.strongly like	2,4	5,0	2,5	7,5
Total	100	100	100	100
Average	3,84	4,13	2,44	3,08
N	124	119	119	120

Figure 16: Students' opinion upon 4 types of General Implicit Solutions

The reluctance for LACK of data is in accordance with the low rate of success for this kind of questions. On the opposite, the (lowest) reluctance for ABSURDITY contrasts with the fact (insufficiently known by the learners) that this kind of question is the easiest one, i.e. elicits the highest rate of success.

6.8 Preferences for paper or computer administered MCQ SGI tests (N=119)

Computer administered	1	31,1 %
	2	11,8 %
	3	5,0 %
	4	13,4 %
	5	9,2 %
	6	13,4 %
Paper administered	7	16,0 %

Again, responses are spread, with an advantage to computer adminstered test.

6.9 Comments on advantages and drawbacks of MCQs GISs

The end of the questionnaire asked for the major advantages and drawbacks of
- MCQ GISs with open books followed by collective correction;
- Computer administered testing.

   1° MCQs GISs with open books:

Main drawbacks	Main advantages
Ambiguous questions forcing to look systematically for traps and implying a perfect understanding of written language. (25 %)	Addresses reflexion and comprehension instead of rote learning, and leads to a better mastery of the content. (32 %)
Increased attention, concentration, stress, tiredness. (20 %)	Immediate feedback : score is known the same day. (8 %)
Confusion among GISs : 6-7-8-9 ??? (20 %)	The type of requested answer is clearly specified and the student is not anxious to have "enough" answered. (5 %)
Others : less than 5 %.	Others : less than 5 %

   2° Computer administered testing

Main drawbacks	Main advantages
Impossibility to come back to a previously answered question, no overview of the whole set of questions. (33 %)	Immediate reception of feedback about correctness of responses. (44 %)
Being informed immediately of one's errors and lost of points has a negative effect on motivation. (14 %)	It frees successful students from being examined an other time on the same content. (11 %)
Fear of key stroke failures, computer breakdown, etc. (10 %)	Allows to practice, to see whether the way of studying is correct. (6 %)
Increased attention, concentration, stress. (7 %)	Allows justifications. (6 %)
	Lowers the fear of failure in front of a human person. (6 %)
	Introduces variety, ludic behaviour, change in learning. (6 %)
	Saves time (important in exam period). (5 %)

The interviewed students have been presented the questions only once. This explains the main drawback (impossibility to come back). A TWO RUNS version has been introduced afterwards, on the basis of their opinion.

   The possibility of not being presented directly one's score (and, of course, the correct answer) has also been introduced as a consequence of the second main drawback (negative effect on motivation).

The actual decisions made by the testees are now recorded (time spent on each question during the first and the second run, same for answers and confidence degrees) and analysed (stability in correct answer; good move, i.e. change an incorrect answer into a correct one; bad move, i.e. the contrary; stability in error; changing an error for an other one, etc.). We hope this will enable us to detect cognitive strategies and to provide useful feedback on metacognitive processes.

7 The teacher's role and opinion

7.1 The teacher has to spend time on:

a) Defining objectives (they can actually be contents: sections of a chapter, concepts and subconcepts, etc.)
b) Designing MCQs (with GIS) within these categories of contents and with an appropriate feedback that will appear on the screen.
c) Reading learners' written comments accompanying their answers (NB: when the answer is correct, it has been announced to the students that the learner's (handwritten) comment will not be read at all by the teacher so that this process can only be beneficial to the student) and change the student's answer to the correct one (if comments are convincing).
d) Change the item so that the same confusion cannot happen further.
e) When d (changes in students' responses) is finished, introduce (through keyboard or optical reader) the answers within a computer program (CERT).
f) Analyse item characteristics such as rate of selection of the various alternatives, their discrimination index and (on the basis of this last index), discard some items.
g) Compute the students' score only on the remaining (validated) items.

7.2 Difficulties and advantages

Operations a and b described above are time consuming, but they result in a highly coherent course in which students know in advance very well what the criteria will be (the objectives are given to them) and in which they receive a detailed feedback from assessment procedures, facilitating the remediation process.
Operation d is time consuming too, but it results in an item bank constantly increasing in quality (teachers need cumulating their expertise) on experimental grounds.

Operation c and f can make difficulties for teachers to accept that the quality of their questions can be disputed, but students dislike even more teachers who do not accept any remark! Which teacher can insure that his/her questions are always perfect?

Conscious of the intent of fairness and of the teacher's good will, the students have also the feeling of participating in the process, i.e. that their comments are actually taken into consideration and can have an influence on the system itself.

7.3 New adepts

The TASTE CHECK approach has recently been used (November 92 to January 93) with the (250) graduate students of the University of Liège who prepare themselves in becoming secondary highschool teachers. They followed Leclercq's course on Educational Psychology and were originated from all faculties (from romanists to veterinarians). Their advices are being collected.

Professor Born, Leclercq's colleague at the University of Liège, has engaged the whole process for his course about developmental psychology for (200) freshmen students in psychology.

A.C. Nizet, who assisted Prof. Born, has proposed intelligence and cognitive style tests to these students. This research is in development. A key concern of Prof. Born was that all the students will anyway be assessed in the oral mode, but the interview of excellent students (on the basis of their TASTE scores) can be facilitated (and accelerated).

8 General discussion

The computer administration of GIS MCQs (here with CHECK) presents problems. One of them is the duration of testing. Usually, we allow 90 min for 15 questions, including the time needed by the student to read the feedback displayed by CHECK after each answer. Therefore, some students have not time to see all the questions two times and they are therefore scored on the basis of their first answers to some questions.

General Implicit Solutions (GIS) are only a first step in a more sophisticated way of questioning. In a next version, students who indicate "Absurdity" will be asked to point (with the mouse, or in a written way) where the absurdity is. The same for the answer "Missing - lack of data": they will be asked to specify which kind of information they would like to obtain; they will be given it and invited to answer a second time. The same for the "Reject-None" answer: the student will be invited to type *his/her* correct answer.

As can be seen, the barriers between multiple choice and open ended questions are slightly disappearing, as well as the barriers between training (practicing) and evaluation (scoring).

The student will be more and more placed in a professional situation, i.e. to process "cases". He/she will be more and more in the same position as a physician who receives a series of individual cases in his consultation, detecting "usual" situations, discarding "absurd" diagnosis (often coming from the patients themselves), obtaining more information (inspection of the body, analysis of blood or urine, x-rays) in case of lack of data, detecting multiple disease situations, etc.

It is well known that medical experience comes from not only the exposition (in a passive way) to numerous cases, but essentially from the actual confrontation with cases in a problem-solving situtation.

MCQs with SGI, confidence degrees, the CHECK interactive approach and the TASTE open books principles are steps into this direction: learning by living, even if it is in a simulated way, since education is bound to accelerate the process of learning. How fast it can be accelerated not only on specific contents, but on the enquiry process.

The answer to this question is not around the corner, but seems now less unreachable. We even have the conviction that the whole approach described here is a good direction to be followed and deepened to approach this challenging goal.

References

Barras, H. (1992), QCM sur Minitel en EAD: vers l'automatisation, in Weber & Dumont "Les questionnaires automatisables", Marne-la-Vallée: Colloque de l'ESIEE

Boxus, E. (1988a), Les QCM à solutions générales au service de l'évaluation à livre ouvert, in *Actes du Colloque International "Formation, Evaluation, Sélection par Questionnaires Fermés"*, Marne-la-Vallée: ESIEE, vol. 1, 318-331

Boxus, E. (1988b), Vers des examens à livre(s) ouvert(s)? In *Le défi pédagogique de l'enseignement supérieur*, Actes du Congrés de l'Association Internationale de *Pédagogie Universitaire*, Montréal: AIPU, 533-537

Boxus, E. (1992), Check: une banque de questions interactives, in Weber & Dumont: questionnements automatisables, Colloque International ESIEE: Marne-la-Vallée, 29-49

Boxus, E., Leclercq, D. Osterrieth, S., Wuidar, H. (1991), Principes communs pour évaluer les résultats cognitifs de la formation, Bruxelles: Commission des Communautes européennes, EUROTECNET

Bruno, J. & Baxter, J. (1989), An application of information reference testing. Proceedings of the Sixth International Conference on Technology and Education, Orlando, vol. 2, 191-192

Bruno, J.E. (1987), Using MCW-APM test scoring to evaluate economics curricula. Journal of Computer Based Instruction

Bruno J.E. (1990), Confidence contour tests item analysis with information referenced testing. Paper presented at the Seventh International Conference on Technology and Education, Brussels

Bruno, J.E., Baker, J.B. (1989), Computer assisted learning through technology based classroom formative evaluation: an application of information referenced testing. Proceedings of the Sixth International Conference on Technology and Education, vol. 2, Orlando, Florida

Bruno, J.E. (1993), Using testing to provide feedback to support instruction: a reexamination of the role of assessment in educational organizations. In Leclercq & Bruno (eds.), Item banking: interactive testing and self-assessment. NATO ASI Series F, Vol. 112. Berlin: Springer-Verlag (this volume)

Choppin, B. (1970), An IEA study of guessing. A proposal. Stockholm, International Association for the Evaluation of Educational Achievement. Unpublished memorandum, IEA/TR/9

De Finetti, B. (1965), Methods for discriminating levels of partial knowledge concerning a test item. Brit. J. of Mathem. and Stat. Psych. 18, 87-123

De Landsheere, G. (1984), *Evaluation continue et examens, Précis de Docimologie*, Bruxelles: Labor

Depover, C. (1987), *L'ordinateur média d'enseignement*, Bruxelles: De Boeck

Descartes, R. (1628), Règles pour la conduite de l'esprit

Dressel, P. & Schmid, J. (1953), Some modifications of the multiple-choice items. Educational and Psychological Measurement 13, 574-595

Dudley, H.A.F. (1973), Multiple-choice tests. Lancet 2, 195

Dudycha, A.L. & Carpenter, J.B. (1973), Effects of item format on item discrimination and difficulty. J. Application Psychology 58, 11-121

Dumont, B., Lazerges, J.M. (1992), Des QCM télématiques en mathématiques pour préparer un concours d'entrée dans une école d'ingénieur, in Weber & Dumont, "Les questionnaires Automatisables", Marne-la-Vallée : Colloque de l'ESIEE

Fabre, J.M. (1975), *Docimologie expérimentale et évaluation par questionnaires : étude du jugement multiple et de l'autopondération*, Thèse de doctorat de troisième cycle en psychologie, Université de Provence, ronéotypé

Fabre, J.M. et Noizet, G. (1977a), Confiance attachée aux réponses à des questions à choix multiple, *Journal de Psychologie Normale et Pathologique*, 74, 335-362

Halleux-Hendrick, J. (1969b), Construction des questions à choix multiple: une seule solution correcte?, in *Revue Belge de Psychologie et de Pédagogie*, Tome XXXI, n°127, 113-125, Bruxelles

Hughes, H. & Trimble, E. (1965), The use of complex alternatives in multiple-choice items. Education and Psychology Measurement 25, 1

Hunt, D.P. (1993), Human self-assessment - theory and application to learning and testing. In Leclercq & Bruno (eds.), Item Banking: Interactive Testing and Self-Assessment. NATO ASI Series F, Vol. 112. Berlin: Springer-Verlag (this volume)

Kagan, J., Impulsive and reflective children: significance of conceptual tempo. In J. Krumboltz (ed.), Learning and the Educational Process. Chicago: Rand McNally, 133-161

Karraker, R.J. (1967), Knowledge of results and incorrect recall of plausible multiplechoice alternatives. J. Education Psychology 58, 11-14

Keller, F. & Sherman, G. (1974), The Keller Plan Handbook. Menlo Park, CA: W.H. Benjamin

Keller, F. (1968), Goodbye teacher. J. Applied Behaviour Analysis 1, 78-89

Lambert, F. (1992), Evaluation interactive informatisée et auto-évaluation, graduation Dissertation, Faculty of Psychology and Education, University of Liège

Leclercq, D. (1980), Computerised tailored testing: structured and calibrated item banks for summative and formation evaluation. European Journal of Education 15(3), 521-260

Leclercq, D. (1983), Confidence marking, its use in testing. In B. Choppin and N. Postlethwaite (eds.), Evaluation Education, an International Review Series, 6(2), 161-287n

Leclercq, D. (1986), *La conception des Questions à Choix Multiple*, Bruxelles: Labor.

Leclercq, D. (1987), *Qualité des questions et signification des scores*, Bruxelles: Labor

Leclercq, D. (mars 1988), Mesurer la connaissance partielle et le réalisme par les degré de certitude, in Actes du Colloque International "Formation, Evaluation, Sélection par Questionnaires Fermés", Marne-la-Vallée, vol. 1, 306-316

Leclercq, D. & De Brogniez, P. (1990), A fresh look on confidence marking. In Estes, Heene and Leclercq (eds.), New pathways to learning through educational technology. Proceedings of the Seventh International Conference on Technology and Education (ICTE), Brussels, March 1990, vol. 1, 646-649

Lindley, D. (1971), Making Decisions. London: Wiley

Luce, R. & Raiffa, H. (1966), Games and Decision. New York: Wiley

Luce, R. (1959), Individual Choice Behavior. New York: Wiley

Massengill, H. & Shufford, E. (1967), What pupils and teachers should know about guessing. Technical Report SMC R-7, Lexington, MA

Noizet, G. & Caverni, J.P. (1978), *Psychologie de l'évaluation scolaire*, Paris: PUF

Noizet, G. & Fabre, J.M. (1975), Etude docimologique des questionnaires à choix multiple (QCM) : Perspectives de recherche, *Scientia Peadagogica Experimentalis*, vol. 12, 38-62

Pitz, G.F. (1974), Subjective probability distribution for imperfectly known quantities. In L. W. Gregg (ed.), Knowledge and Cognition. New York: Wiley

Pyrczak, F. (1972), Objective evaluation of the quality of multiple-choice test items designed to measure comprehension of reading passages. Read. Res. Quart. 8, 62-71

Pyrczak, F. (1974), Passage-dependance of items designed to measure the ability to identify the main ideas of paragraphs: Implications for validity. Educational and Psychological Measurement 34, 343-348

Savage, J. (1971), Elicitation of personal probabilities and expectations. J. American Statistical Association 66, 336, 783-801

Shuford, E., Albert & Massengill, N.E. (1966), Admissible probability measurement procedures. Psychometrika. 31, 125-145

Shuford, E. (1993), In pursuit of the fallacy: resurrecting the penalty. In Leclercq & Bruno (eds.), Item Banking: Interactive Testing and Self-Assessment. NATO ASI Series F, Vol. 112. Berlin: Springer-Verlag (this volume)

Slakter, M., Koehler, R. & Hampton, S. (1970), Learning test-wiseness by programmed texts. J. Educational Measurement 7, 247-254

Smith, R. (1970), An empirical investigation of complexity and process in multiple-choice items. J. Educational Measurement 7, 33-41

Strang, H.R. & Rust, J.O. (1973), The effects of immediate knowledge of results and task definition on multiple-choice answering. J. Experimental Education 42, 77-80

Strang, H.R. (1977), The effect of technical and unfamiliar options on guessing on multiple-choice test items. J. Educational Measurement 14, 253-259

Tversky, A. & Kahneman, D. (1974), Judgment under uncertainty: Heuristics and biases. Science 185, 1124-1131

Von Neumann, J. & Morgenstern, O. (1974), Theory of games and economic behaviour. Princeton University Press

Wahlstrom, M. & Boersma, F. (1968), The influence of test-wiseness upon achievement. Educational and Psychological Measurement 28, 413-420

Williamson, M.L. & Hopkins, K.D. (1967), The use of "none of these" vs homogeneous alternatives on multiple-choice tests: Experimental reliability and validity comparisons. J. Educational Measurement 4, 53-58

Wood, R. (1974), Guessing on objective type test items. School Science 56, 179-180

Wood, R. (1976), Inhibiting blind guessing: the effect of instructions. J. Educational Measurement 13, 297-307

Wood, R. (1977), Multiple-choice: A state of the art report. In Choppin & Postlethwaite (eds.), Evaluation in Education International Progress. Oxford: Pergamon

Self-Confidence Assessment During Computer-Assisted Testing in Histology

P. Gathy and J.-F. Denef

Laboratory of Histology, Medical School, Louvain Catholic University,
Avenue E. Mounier, 52, UCL 5229, B-1200 Brussels, Belgium.
Phone: +32-2-7645229, E-mail: denef@isto.ucl.ac.be

Abstract: A computer-assisted self-testing system has been developed allowing students to evaluate their own knowledge and their ability for identifying unknown histological structures. It consists of an image digitizing-editing system, an image database shared through a local area network, and teacher's and student's environments.

The authoring system allows the teacher to define text frames, multiple-choice and open-ended questions. The student's environment includes on-line help and recording of the student's identification, answers, scores and messages. The educational impact of the system was tested with a limited number of students. Depending upon previous evaluations, two groups of good and of weak students were randomly selected, half of whom used the system. Marks in histology at the final examination were significantly increased for the weak students as compared to the subgroup of weak students not using the system. On the other end, no significant difference was observed among the good students.

The system has been implemented with a self-confidence assessment (SCA) procedure introduced before the correction of the student's answer. Up to July 1992, thousands of SCA scores were recorded, analyzed and correlated with the academic results.

A strong positive correlation was found between the SCA scores and the marks at the final examinations for good students whereas a negative but loose correlation was observed for weak students.

Keywords: CAL, Self-assessment, Histology

1 Introduction

At the Catholic University of Louvain Medical School, first-year students have been taught general histology by an audio-visual method since 1972 [1]. The lessons are given in a classroom equipped with 80 individual cubicles, where each student can learn at his own pace, using a slide projector, a tape recorder, and a microscope. Whenever he wants, the student may ask teachers or senior students for explanations. The objectives of the courses are less the acquisition and the factual recall of theoretical knowledge than its integration, management and application in problem solving situations. The course has been designed to promote the way of reasoning very similar to the diagnostic reasoning strategy of physicians.

These problem-solving skills are reinforced weekly through small discussion groups in which students are asked to apply their theoretical knowledge to identify unknown tissues or structures. During this process, each student's factual recall and reasoning skills are evaluated. These evaluations have been proven to be predictive for the final success at the year after only ten weeks of curriculum. Students failing at least three times during the first four lessons have success rates below 10% at the final examinations [2].

The computer-assisted testing system was designed in 1986 to help the students to further train and improve their reasoning strategy. It is individual, self-paced and based on a socratic, mastery-oriented approach, providing the students with a more adequate and active feedback than the teacher can provide with a group of 8 to 10 students. The major aim of this system is neither to transfer factual knowledge (the role of the audio-visual course), nor to perform a formal student's evaluation (one of the teacher's roles), but to help the student to apply his previously acquired knowledge in practical situations. The first tests began with a prototype in October 87.

In this system, a self-confidence assessment procedure was also developed to bring the students to assess the reliability of their answers.

2 Material and methods

2.1 Hardware

The self-testing system has been described in details elsewhere [3]. A first release was running on an IBM personal computer interfaced to a laser disk player (Philips Laservision VP410, Belgium) connected to a colour TV monitor. The laser disk (Erasmus University of Rotterdam, Netherlands) contained about 8000 fixed analogic histological images.

A second release using digitized images was developed on PC (286 or higher) and on IBM personal systems/2 (Model 50 or higher). Digitized images were displayed on enhanced VGA adapters providing a resolution from 640x480 to 1024x768 pixels with 256 colours. Images were produced from a high resolution video camera mounted on a microscope and connected to a colour frame grabber, or from gray scale or colour scanners. They were processed (filtering, contrast enhancement, colour correction), compressed, stored in the TGA file format (AT & T, Truevision Inc, USA) on a file server (IBM PS/2 model 80 with a 311 Mbytes hard disk), and distributed to workstations through a token ring local area network (IBM PCLAN 1.30). A picture editing program was used to add legends or enhance structures on the images. A database program also allowed the retrieval and display of digitized images, according to several indices such as image number, tissue or organ type.

2.2 Educational software

Computer-assisted self-testing software was written in Turbo Pascal 5.5 (Borland, USA). It consists of a teacher's and a learner's environment. It is easy to use and does not require any programming skills.

The *teacher's environment* presents a menu allowing to edit, test or print a session. A session is a sequence of frames, optionally referring to images, and linked to each other by direct or conditional branchings. Four kinds of frames are used: text frames, simple multiple-choice, complex multiple-choice and short open-ended questions. *Text frames* are used to give the student explanations, evaluations or feedback comments, without requiring an answer. *Simple multiple-choice questions* ask the student to type in a single answer, for instance: "true", "false", or "I don't know". *Complex multiple-choice questions* present up to 10 items, among which the student has to point out the relevant ones. *Short open-ended questions* consist of a statement, followed by 15 possible answers, marked correct or incorrect by the teacher and compared to the student's answers to assess his response. Wild card characters can be included by the teacher to match several student's answers even with spelling mistakes. A score is associated with each answer, so that the student's performance is easily evaluated. Three short comments, associated with each frame, are set up for immediate feedback, according to the student's answer. The frame ends up with a branching structure, allowing adaptive testing. If the student's answer is right, he goes further into the session and gets more difficult questions, otherwise, the same question is repeated, but associated with another image. The teacher strictly controls the student's environment. He manages the list of authorized students, defines which sessions are accessible, and creates the log-on messages for the students. These messages are intended either for all the students, for instance to let them know when new sessions are available, or for one or another student, for instance to give him an acknowledgement when he leaves a comment in the system mailbox.

The *student's environment* is user-friendly and self-explanatory. On-line help is always available and current instructions are displayed at the bottom of the screen, so that students can use the computer without any user's guide. Different colours are used to enhance comments, error messages or incorrect responses. At every moment, students can leave comments in the system mailbox. Most often, they use the mailbox to inform the teacher when they do not understand a question, or when they disagree with the correction. This is very helpful to detect mistakes in new sessions.

To sign on into the system, the student has to enter his name and an eight digit identification number. A menu is then presented with the available sessions. After each answer, the correct responses, scores and appropriate comments are automatically displayed. Computerized sessions follow the same guidelines as the weekly discussions with the teachers, access to the next part of a session is not allowed until the student masters the current one. Nevertheless, each student is allowed to repeat a session as many times as he wants. Student's name, session title, time spent, answers and messages are recorded in a text file. This allows the teacher to assess the student's performance and to further improve sessions previously developed. This log file is then split up into summary text files, which in turn are submitted to statistical packages such as SPSS-X (SPSS Inc.,

U.S.A.). This allows to calculate parameters such as the total time spent on the system by student, by session or by workstation.

2.3 Self-confidence assessment procedure

The system allows a self-confidence assessment (SCA) procedure. Its introduction in each step of a session depends on the teacher's will. When implemented, the student is asked, after introducing his answer and before getting it corrected, to define his level of self-confidence. A small message appears at the bottom of the frame, which displays a scale between zero (meaning "I have answered at random, I am not sure at all of my answer") and 100 (meaning "I am absolutely sure"). Until May 92, this scale extended linearly, with 11 levels from 0 to 100%. It was then changed to a progressive scale with 6 levels only: 0, 25, 50, 70, 85, 95, and 100 %, recommended elsewhere [4]. The student gives his level of self-confidence in a graphical way: any hit on the right or left arrow of the keyboard makes a little bar grow or shrink step by step. Until May 92, students were allowed to answer the self-confidence assessment by simply hitting the enter key, the default value 100% thus indicating that they were either completely sure, or unwilling to answer. Since May 92, for 3 months, the system was modified: students were then forced to assess their certitude by hitting at least once the right arrow key.

2.4 Testing the pedagogical impact of the system

The prototype of the system was tested in 1987 by the class of first-year students in Medicine. Two groups, according to the first four evaluations in histology were designed according their previous evaluations. The first group had obtained at least two satisfactory evaluations and had no apparent problem. Students from the second group had at least three unsatisfactory marks and were encountering difficulties in histology. To test the efficiency of the computer-assisted self-assessment system, each group was divided into two subgroups. The first subgroup consisted of 24 students, randomly selected to use the computer. They were asked to freely complete at least one individual session for each of the 7 chapters of the course. The control subgroups consisted of 24 students each, randomly selected *a posteriori* among good and weak students not allowed to use the computer system. The computer was placed in the laboratory rather than in the classroom, so that only authorized students could use it. Each student from the test subgroup received a single leaflet which briefly described the keyboard, and explained how to sign on into the system and how to introduce the answers. Finally, the teacher who performed the final oral examination in histology was unaware of which students were selected for the test. At the end of the test, students who completed individual sessions for less than two chapters were included in a drop-out subgroup (see below).

Next year, 77 students were asked to evaluate some pedagogical and technical aspects of the system in an anonymous questionnaire.

3 Results

3.1 Pedagogical impact

Among the two subgroups of 24 students selected for prototype testing, 12 students did not participate and a further 5 completed individual sessions for one chapter only. These students were separated from the others and included in a drop-out subgroup. Of these 17 students, 3 did not attend the final examination. Another student completed several sessions, but did not attend the final examination, so that each test subgroup consisted of 15 students.

Students completed 602 individual sessions during the test. On average, each session lasted between 10 and 12 minutes and each student spent 3 hours 20 minutes on the computer. Performances at the final oral examination in histology were rated from 0 to 20. To discriminate between good and weak students, a two way analysis of variance was performed, followed by a F-test and by a comparison of means through a two-tailed t-test (Table 1). Among the good students, no significant difference was observed between test, drop-out and control subgroups. In the group of weak students, 3 students from the drop-out subgroup and one from the test subgroup did not attend the final examination. However, those who did use the computer attained significantly higher scores than the control and drop-out subgroups.

Table 1 Mean scores, standard deviation and number of students (final examination in histology)

	First group (Good students)			Second group (Weak students)		
	Mean	S.D.	n	Mean	S.D.	n
Control subgroup	12.6	3.61	24	6.9	4.90	24
Drop-out subgroup	10.1 [a]	5.09	9	3.4 [c]	4.98	5 [e]
Test subgroup	11.9 [b]	4.47	15	10.2 [d]	3.28	15 [f]

[a]	N.S. vs control subgroup
[b]	N.S. vs drop-out and control subgroups
[c]	N.S. vs control subgroup
[d]	$p < 0.05$ vs drop-out and control subgroups
[e]	(3 students did not attend the final examination)
[f]	(One student did not attend the final examination)

Results of the anonymous questionnaire completed by 77 students in December 88 are shown in Table 2. Nearly 90 % of the students appreciated the educational aspects of the system and found it to be helpful in improving their theoretical knowledge and their ability to identify histological structures. Technical aspects were also satisfactory: quality and transfer rate of digitized images through the network were favourably judged. In contrast, very few students thought this system could replace the weekly discussion groups with teachers. Students also wanted more workstations to be made available.

Table 2 Percentage of positive appreciations on the self-assessment system

Usefulness in increasing theoretical knowledge	90
Usefulness in increasing the ability to identify structures	94
Quality of digitized images	92
Interaction with the system	74
Possibility to replace weekly discussions with the teacher	21
Global acceptance of the system	88

3.2 Frequentation

Since October 89, when workstations were placed in the microcomputer room of the Faculty library, the monthly time spent on the system has constantly increased, up to 510 hours in May 90. Most of this time (76%) was on the workstations located in the microcomputer room. From October 87 to May 90, 620 students have used the system, for a total of nearly 3000 hours. On average, each student has completed 24 individual sessions, lasting nearly 4 hours 30 minutes.

During the academic year 90-91, 479 students (out of about 600) spent a total of 4282 hours on the computers, using the system for 18059 sessions. The mean time per student was 6h20 for 29 sessions.

3.3 Self-confidence assessment

Almost all teachers implemented the SCA procedure for almost all questions. Only one teacher did not, considering it not useful. The SCA scores were analyzed separately each year. All the calculations were based on the upper limit of each self-confidence score.

In 1987, the scores were high and no significant difference was found between good and weak students. This year, as for the following years, a strong positive correlation ($p<0.001$) was found between the SCA scores and the results at the final examination in histology for good students but not for weak students.

The following year, the positive correlation for strong students was confirmed, while a negative although less significant ($p = 0.03$) was found for weak students.

From the year 1992, results obtained with SCA scores tested with the linear or the progressive scales were compared. No significant difference was found so far.

As a rule, students over-estimated their performances by about 10-15%. This over-estimation varied depending on the chapter analyzed but also among different sessions on the same chapter. This suggests that the style of the questions and the structure of the session itself may influence the self-confidence scores of the students.

4 Discussion

4.1 Effectiveness of a computer-based learning system

The evaluation of the effectiveness of a computer-based learning system should include a measure of the time spent by the teacher preparing lessons. Times reported vary strikingly among authors between 10 to 100 hours to prepare an hour of instruction [5, 6]. Many authors emphasized the usefulness of authoring systems in reducing the time spent in creating tutorials without requiring programming skills [7, 8]. Our system was designed as an authoring system including an editor, providing thus a framework for each session, so that the teacher could focus on the educational content. To create a work session, time spent by the teachers at the terminal was estimated at 3 to 4 hours, whereas design and elaboration of digitized images took between 10 and 15 hours for a work session lasting between 10 and 20 minutes. However, since the framework of one session can be reused in another one, the total time required tended to decrease.

The improvement of student's performances by computer-based instruction has been questioned by Marion and colleagues[9] and Jacoby and colleagues [6]. Almost all studies established that for the same performances, students spent less time with computer-based instruction than with traditional lectures. Woods and colleagues [10] suggested that low-scoring students principally benefit from computer-assisted instruction. Becker [11] stressed that computers affect differently lower-ability students and tend to increase their motivation, self-confidence and self-discipline. Our results confirm these findings, since a significant increase in marks at the final oral examination was observed only for academically weak students. The relatively large number of students who dropped out of the test might be due to a lack of motivation or to the overload of the sole workstation available during prototype testing. This drop-out rate should have artificially increased the average marks of the test subgroups, by excluding less motivated students. However, the drop-out rate was the same for the good students as well as for the weak ones. Furthermore, the final examination showed no significant difference between the drop-out and the control subgroups. It also seems unlikely that the difference observed between test and control subgroups for the weak students might be simply due to a difference in motivation. The possibility for the weak students to train repeatedly without the direct intervention or the judgement of the teacher might also have increased their self-confidence. Also, students seem to prefer working with the library microcomputers rather than with the histology classroom microcomputers: they feel more free and less controlled. Furthermore, they can easier discuss together in the library, where there is more space in front of each microcomputer. One should also keep in mind that this system has been introduced as an additional learning medium to the audio-visual course, which should increase the self-directed learning skills of the students. Students themselves saw this system as a useful additional learning tool.

4.2 Some practical considerations

Although we have been collecting data for several years now, practical changes in the availability of the computers and the behaviour of the students make the data less and less reliable. Indeed, during the first year, the access to the PC was controlled. During the second year, the computers were located in cubicles, impairing groups of more than one student per PC. Later on, and mainly in the library, we observed that students worked by groups of 2-4, discussing together before giving an answer. Thus, the self-confidence score is no longer the measurement of a single learner, but rather the evaluation of a small group. Of course, this impairs the study of correlations with examination results.

However, we did not change the computer access for two reasons. Since we have too few workstations, this allows a larger number of students to work on the same time. We are also convinced, although this should probably be demonstrated, that the discussions occurring between students do help them in their learning process.

4.3 Self-confidence assessment

The results of the SCA procedure are still preliminary and further analysis is needed to evaluate its impact. In our opinion, their interpretation largely depends on the fact that the self-assessment performed in our conditions is formative and not normative. Further studies, which are still in progress, tend to prove that weak students almost always over-estimate their ability to answer basic questions in histology. This could at least partially explain the negative correlation observed among weak students between SCA scores and the final examination in histology. Nevertheless, several questions can be raised :

- Does a self-confidence assessment improve the learning process among our students? Would an effective pay-off system increase its impact? Is it clear that SCA provides informations for the teacher? Does it directly alter the behaviour of the students?

- Is there any variation of the SCA scores depending on the time, for example if students repeat the same work session, or perform another session dealing with the same subjects?

- Why do we observe differences between the SCA scores and the performances in various sessions, written by different teachers?

These are questions we would like to address in the near future.

References

Abdulla, A. M., Watkins, L. O., Henke J. S.: The use of natural language entry and laser videodisk technology in CAI. J. Med. Educ. 59, 739-745 (1984)

Becker, H. J.: Using computers for instruction. Byte 12, No. 2, 149-162 (1987)

Gathy, P., Denef, J-F., Haumont, S.: Computer-assisted self-assessment (CASA) in histology. Computers Educ. 17, 109-116 (1991)

Haumont, S., Cordier, A., Dambrain, R., Herveg, J-P., Rousseau, P., Sprumont, P.: A new approach to teaching histology at the University of Louvain. Med. Biol. Illus. 24, 202-206 (1974)

Haumont, S., Denef, J-F., Gathy P.: Enseignement audiovisuel et évaluation continue en histologie. Rev. Educ. Méd. (Paris) 11, No. 2, 45-49 (1988)

Jacoby, C. G., Smith, W. L., Albanese, M.A.: An evaluation of computer-assisted instruction in radiology. A. J. R. 143, 675-677 (1984)

Leclercq, D., Boxus, E., De Brogniez, P., Lambert, F.: The TASTE approach: General implicit solutions in MCQs, confidence marking, open books exams, and interactive testing. In: D. Leclercq, J. Bruno (eds.) Item banking: Interactive testing and self-assessment. NATO ASI Series F, Vol. 112. Berlin: Springer-Verlag (this volume)

Marion, R., Niebuhr, B. R., Petrusa, E. R., Weinholtz D.: Computer-based instruction in basic medical science education. J. Med. Educ. 57, 521-526 (1982)

Miller, K., Willett, J. E.: Audio-enhanced computer-assisted learning in science teaching. Biochem. Soc. Trans. 15, No. 3, 351-354 (1987)

Smith, S. G., Sherwood, B. A.: Educational uses of the Plato computer system. Science 192, 344-352 (1976)

Woods, J. W., Jones, R. R., Schoultz, T. W., Kuenz, M., Moore R. L.: Teaching pathology in the 21st century: An experimental automated curriculum delivery system for basic pathology. Arch. Pathol. Lab. Med. 112, 852-856 (1988)

Distance Interactive Testing

Bernard Dumont

Université Paris 7 - 2, place Jussieu, 75251 Paris Cedex 05, France
Tel. 33-1-44 27 60 74, Fax 33-1-44 27 57 40

1. Self-Testing in Distance Learning

Learning at a distance involves some particular difficulties because of the delay between learning and correcting. On classical correspondence courses, learners can wait for several weeks (or months sometimes!) before getting their homework back (Fig. 1). This situation is obviously a source of instability and can partly explain the high dropout rate in classical distance education systems.

In the first period, the student is trying to understand and to assimilate some concepts. At the end of one or more periods, he sends his homework to the institution. Then, he goes on a next chapter, re-using, in general, what he had learnt in the former periods. When he receives his homework back, 3 or 4 weeks later, usually he will find a lot of mistakes or misunderstood notions.

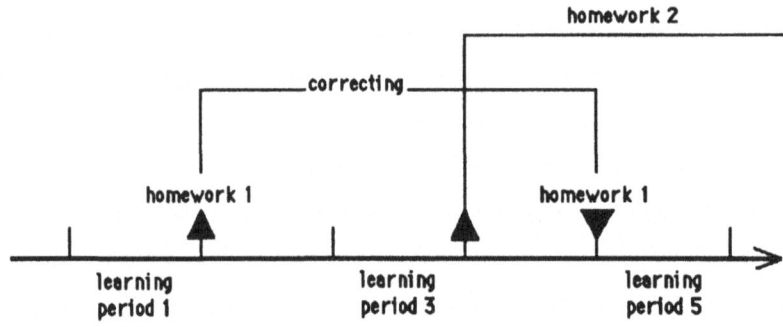

Figure 1: Time for correcting in correspondence education

The problem is then very crucial: the student has to go back to the mental situation he was at the time he wrote his homework. He must understand why he missed some points or misunderstood others. But, until then, he believed, deeply or not, he was OK with these points. The problem is even harder because of the moment he received this information: he has to rethink about all what he learnt since this moment, to "rebuild" the already learnt material, taking into account the new information he has now at his disposal. He can be very anxious too, if he has already sent another homework for correcting.

This organization requires a lot of time and concentration from the student who can easily be disturbed and be psychologically weakened. So we are proposing a kind of organization, based on a "pedagogy of errors" and a self-evaluation testing material[1].

1.1 Didactical bases for a "pedagogy of errors"

Errors are part of the normal learning process, even if our Judeo-Christianized tradition is identifying "error" with "fault". The main idea is to emphasize work on errors as a source of information about misunderstanding (Dumont 1989). In a face-to-face teaching mode, these errors are sources of information for the teacher who is able, analysing them, to understand in which ways his students are assimilating or not the new concepts. A good teacher can be characterized by a great facility to modify his lecture depending on the reactions and errors of his students, even if often it is almost unconsciously. He must improvise, in a positive meaning of the word. In a distance course, student is more often alone with his problems, and the text book is not able to vary depending on the way it is read! Another problem: it is even more dramatic for a student to believe he has understood rather to be stopped because he does not understand. In a distance learning process, it is necessary to prevent such mistakes, and to prepare tools for self-evaluation and instant repairing.

1.2 Tools for a "pedagogy of errors"

To give the students means of recognizing misunderstandings, we need exercises with the following capacities:

a) they have to be incitative to make errors: in self-evaluation, it is not negative to make mistake, but it is dangerous not to see where are the problems! These exercises must reveal hidden misunderstandings;

b) wrong answers must be easily analyzed with some specific grid, helping student to correctly diagnose the origin of the error;

c) question formulation must not interfer with the way the student is answering:
   - student should not answer "correctly" with incorrect reasoning,
   - student should not be mislead to a wrong answer beacause of linguistic biases.

We recommend that such exercises be built upon the results of a didactical research program.

To achieve point a), we need to know what kind of errors (and in which context) errors on a specific matter are made. A good way to obtain this didactical

---

[1] This kind of organization does not deny the interest of homework being corrected, but I think it is better for the learner to have less homeworks to write. So he will be able to concentrate on his learning and to evaluate by himself, as soon as possible, what he has just learnt. Homeworks arriving after a longer period of learning, they will be an opportunity of synthesis, that is more difficult to self-evaluate.

knowledge, is to make large "paper and pencil" inquiries among sample of the target population. (cf. our research on logic or on fractions upon several thousands of people; de Géry et al. 1985, Dumont 1989, Dumont et al. 1990, Dumont 1991). In that way, it is possible to observe the influence of different parameters (question formulations, numerical values,...) using several variants of the same test, and to build databases of errors with statistical hierarchy of appearence.

On point b), we have created original MCQ (multiple choice questionnaires), which are presented below.

2. MCQ in Mathematics

From 1983, ESIEE, a school of engineers in Paris (now in Marne-la-Vallée), is using MCQ in Mathematics and Physics to select students (less than 200 among 1500 candidates).

Each test is built with 24 questions. Each question has 5 items For each item, the candidates must choose between no answer, TRUE (T) and FALSE (F). In fact, each item is built to have a unique value: true or false. The number of answers "TRUE" in a question can be any integer between 0 and 5, with no previous indication on this number (Fig. 2).

Number of T	0	1	2	3	4	5
N of exercises	1	3	9	7	3	1

Figure 2: Distribution of "TRUE" in the 24 questions. Concourse 1992

This possibility presents some important advantages:

a) *Non-answer and answer "NO"*

In a classical MCQ where boxes have to be checked to show the "right answer(s)", the fact a box is not checked can be interpreted, at least, in 3 different ways:

i) the student believes the correct answer is not to check the box,
ii) the student understands the question, but he does not know if he must check the box or not, in doubt, he does not check it,
iii) the student has not enough time to answer this question or does not understand the question, so he does not answer.

This ambiguity is a real difficulty to interprete the result when the correct answer is not to check the box: in case (a), the student is right and he knows why, but in the other cases we could give the student a good mark (for the concourse) or wrongly say "Bravo!" in a self-evaluation. With our system, case (a) will be a T, (b) a F and (c) an abstention, without ambiguity.

b) *Coherent answer*

Our exercises are built in a way a same notion is tested in different ways through a combination of some of the 5 items to test the coherence (or consistence) of the student's answer.

For example, let us imagine the following exercise:

	correct answers	Tom's answers
item A:	T	T
item B:	F	F
item C:	F	T
item D:	T	T
item E:	F	F

A classical way of correcting is to mark each item independantly of the others. But if we want to minimize the risk of accepting as a correct answer a "good" choice resulting from a wrong reasoning, we must have other ways to verify not only the final choice but the "global answer". In fact, we can say Tom seems to have succeeded in 4 items among 5. But, if we have a logical link between item A and item C, as:

if somebody succeeds in A and fails in C, then he has probably made a wrong reasoning, answering item A.

In a selection situation, we can penalize Tom instead of encourage him. In a self-evaluation mode, we can reveal him a misunderstanding process and help him to understand better the origin of his wrong reasoning.

Presently, candidates answer on optical papers and the marking is made by computer from a list of boolean evaluation equations (Fig. 3) where logical incoherent answers are taken into account: positive mark due to "correct" answer are compensated by negative ones due to the incoherent global answer. We have not yet made logical links between different questions of the concourse, but it would be very interesting to try.

Candidates are aware of the way we mark copies. It is better not to answer than taking the risk of an illogical global answer.

With this system, if a candidate answers at random, for every question (of the 24 questions of the test) he has 2^5 (=32) possible answers. We have simulated about 1500 candidates answering at random (without blank answers): none of them would have been selected to enter ESIEE (with our grid including penalty for illogical global answer).

EXERCISE

$-\dfrac{\pi}{6}$

$\displaystyle\int \tan x \, dx$ is equal to: (A) $\dfrac{4}{3}$ (B) $\dfrac{2\sqrt{3}}{3}$

$\dfrac{\pi}{3}$

(C) $\dfrac{1}{2}\ln 3$ (D) $\ln\sqrt{3} - 2\ln 2$

$\dfrac{\pi}{6}$

(E) $\displaystyle\int \tan x \, dx$

$\dfrac{\pi}{3}$

EVALUATION GRID

equation	mark
A = " "	-0,5
B = " "	-0,5
C = " "	-0,5
D = " "	-0,5
E = " "	-0,5
B = " F "	+ 1
C = " T "	+ 3
D = " F "	+ 1
E = " T "	+ 5
A = " T "	-10
C = " F "	- 5
D = " T "	- 5
E = " F "	- 3
B = " T "	- 10
A = " T " AND (B=" T " OR C=" T " OR D=" T ")	- 10
B = " T " AND (C=" T " OR D=" T ")	- 10
C = " T " AND D="V"	- 10

Fig. 3: Example of exercise with its evaluation equations

3 Telematics MCQ Item Bank

A research has been organized at the French National Institute for Pedagogical Research, between 1987 and 1990, with collaboration of ESIEE and University Paris 7, the objective of which was to put the tests of ESIEE as a database on a telematics self-evaluation service. A lot of difficulties have resulted from the technical characteristics of Minitel.

A survey of this research is presented in (Dumont and Lazerges 1992), where several services offering MCQ in Mathematics are evaluated.

3.1 Presentation of some MCQ in Mathematics on Minitel

Almost all of these services are built as "one choice among 3 or 4 choices" and they do not give any information on the answer of the user; only "right" or "false". Only one service, called COGITO, gives the user, after an incorrect answer, an associated message for help (cf. Figs. 4a,b) and allows to answer again (but there are only 3 choices).

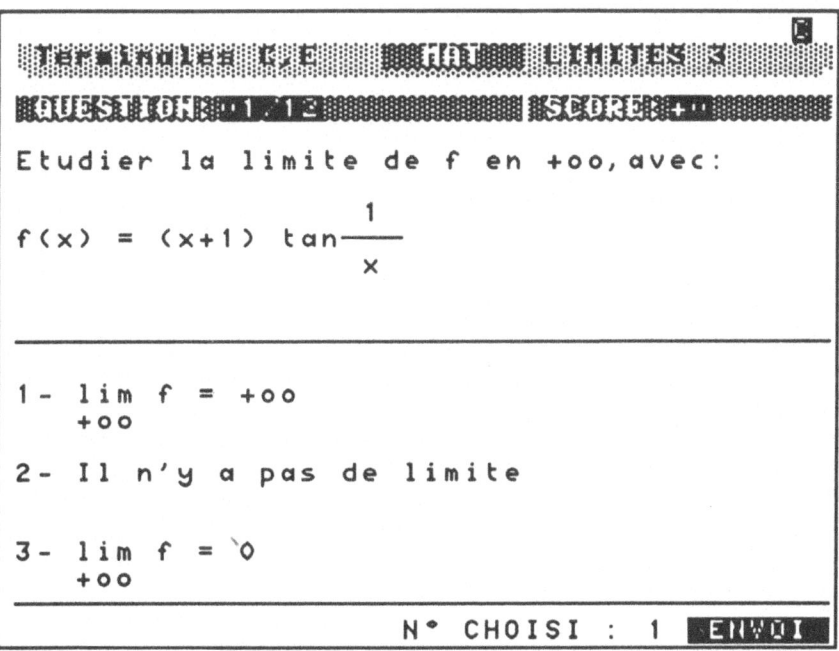

Figure 4a: Exercise from COGITO

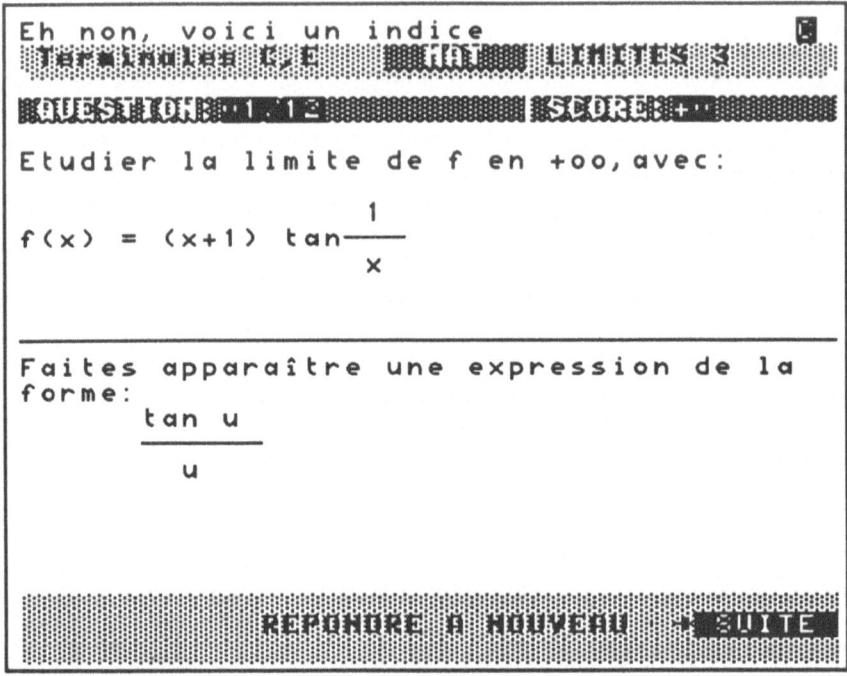

Figure 4b: Exercise from COGITO

3.2 INRP-ESIEE-University Paris 7 Project

The base includes 30 exercises from several ESIEE concourses. Its first interest is the fact we have used statistics from the real test with the answers from more than 1500 students.

The second originality is the kind of help: the answer is analysed globally for a question, not item by item. So it is possible to give the user some information on his error or on an illogical aspect of his answer. At any time, the user can obtain some different levels of help: on one hand, there are between 1 and 6 screens from general information to specific remarks, without giving the answer; on the other hand, several screens giving the complete explanation of the question (Fig. 5).

An answer is a sequence of 5 terms with V ("vrai" for TRUE), F ("faux" for FALSE) or - for blank.

For example: $\boxed{\text{V V F V F}}$ ou $\boxed{\text{V - - V F}}$.

If the answer is utterly correct, the student is congratulated but also invited to look at the explanation of the solution.

If not, the student's answer is compared with the answers given by at least 10% of the candidates having gone through the corresponding examination. In these cases, a specific sequence of screens is send to the Minitel, for help and repairing.

Some illogical patterns are tested. They look like $\boxed{?V??V}$ where ? is used as a joker: the computer does not take into account any answer values at these places.

If there is no similarity between the student's answer and theses filters, the analysis continues item by item.

The real particularity of this database is to help students to work from their own errors. It is as a distance CAL program based on a pedagogy of errors in Mathematics.

Nevertheless, Minitel contraints about writing mathematical formulæ (cf. Fig. 5), are too important and require from the authors too much to express themselves (and too much time to create screens). At the other end, users need to "translate" or to interpret what appears on the screen, which is sometimes very far from what they are used to see on their handbooks.

The future is not obviously in this kind of small terminal without graphics, but anyway the work done here proves the possibility and the interest of creating self-evaluation MCQ on a serious and rich didactical research.

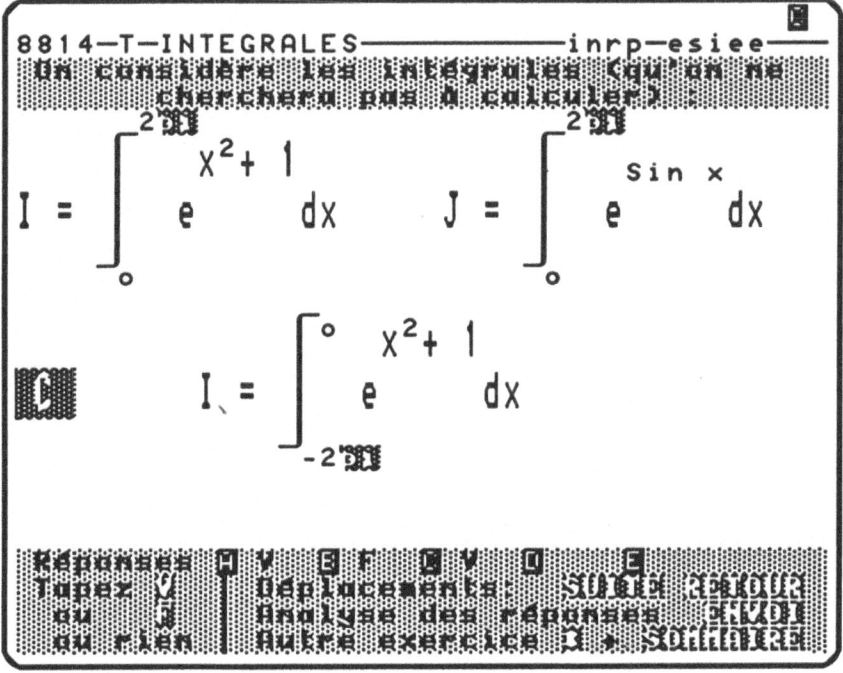

Figure 5: Example of screen with mathematical formulæ

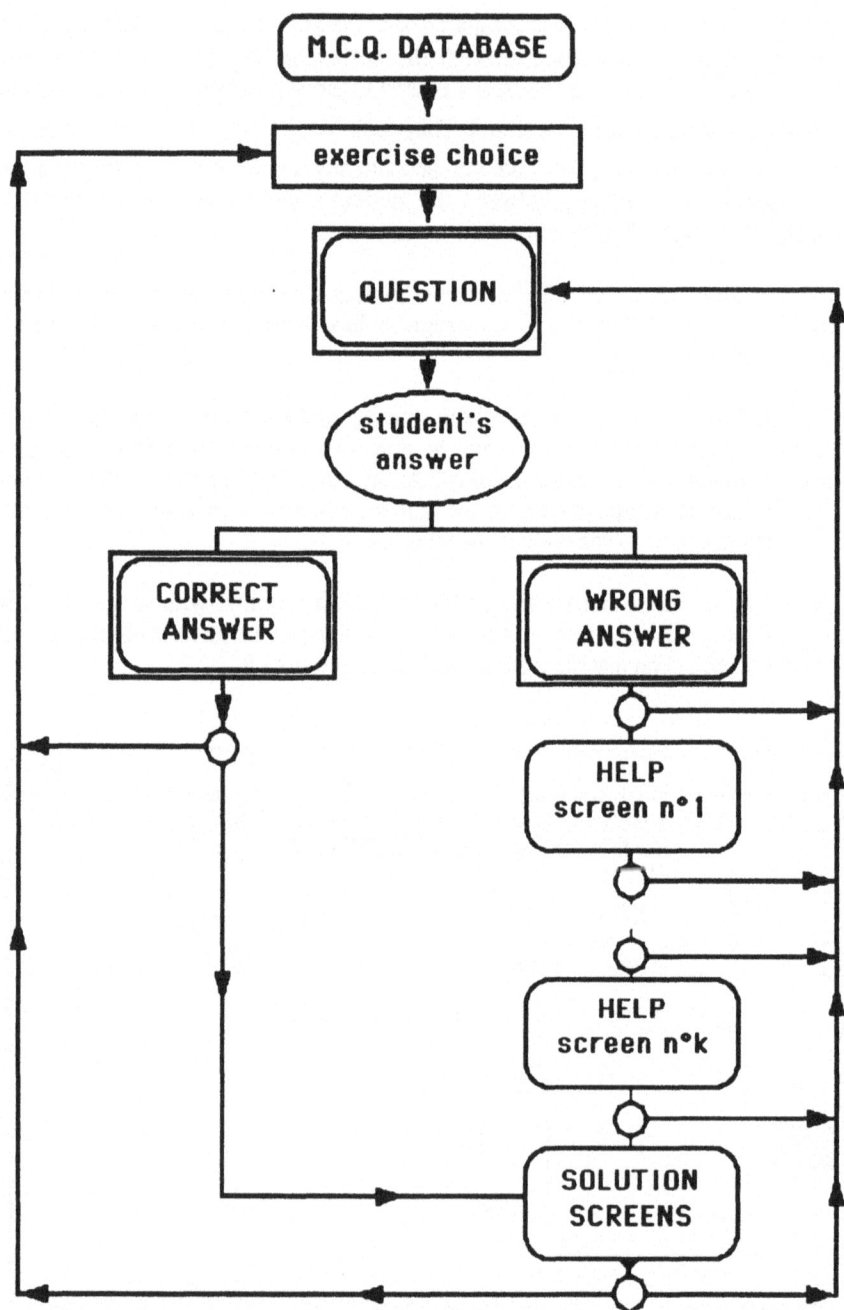

Figure 6: Organization of the service

References

Chastenet de Géry, J., Drouhard, J.-P., Dumont, B., Hocquenghem, S., Lacombe, D., Sol, G., Computer assisted testing by questions and answers. Proceedings of the 4th World Conference on Computers in Education, Part 1, North-Holland, 1985, pp. 311-318

Dumont, B., Questionnements et interprétation des erreurs en mathématiques (Elaboration de modèles pour la compréhension des comportements de réponse et la construction d'outils pédagogiques à supports technologiques), Thèse de Doctorat d'Etat ès Sciences, Université Paris 7, Paris, 1989,

Dumont, B., Drouhard, J.-P. et Grugeon, B., FRACT: un exemple d'apports possibles de l'intelligence artificielle à la pédagogie de l'erreur en mathématiques". in Actes des Journées "E.I.A.O. du PRC-GDR-IA, (ENS Cachan, 18-19 décembre 1989), Institut Blaise Pascal, Université Paris 6, CNRS, Université Paris 7, Rapport n° 31/90, décembre 1990, pp. 191-200

Dumont, B., FRACT: A student's error knowledge based system for diagnosis in fraction calculus. In R. Lewis and S. Otsuki (ed.), Advances in research in computers in education. IFIP–North-Holland, 1991, pp. 89-92

Dumont, B., Lazerges, J.M., Des QCM télématiques en mathématiques pour préparer un concours d'entrée dans une école d'ingénieurs. in Actes du 2d Colloque International sur les questionnaires automatisables, ESIEE, Marne-la-Vallée, mars 1992

The Impact of Interactive Computer Based Instruction with Digitized Audio on Low Literate Adult Learners

John A. Gretes and Michael Green

Department of Curriculum & Instruction, College of Education, University of North Carolina at Charlotte, Charlotte, North Carolina 28223, U.S.A.

Abstract: This paper is an account of the design, development and evaluation of the interactive CD-ROM reading courseware (The READY Course) on low-literate adults. The paper describes the content and design of the READY Course and presents evaluation field test findings in the areas of validity and reliability, and it includes information based on a pre- and posttest design with experimental and control groups. Statistical analysis of reading improvement for the experimental and control groups is provided. Possible future implications for the development of this type of courseware are discussed.

Keywords: Adult literacy, Reading, Low-literate adults, Interactive computer assisted instruction

1 The READY Course Overview

1.1 Introduction

The primary effort of Project READY (Reading to Educate And Develop Yourself) involved designing adult literacy instructional software using new technologies that integrated interactive videodisc, digital audio, and CD-ROM. The READY Course has the following theoretical advantages over traditional classroom reading instruction for adults: (1) improved individualization, (2) interactivity, (3) high adult interest level topics, and (4) digitized audio support.

The READY Course is a series of ten multimedia reading instruction modules designed for adults reading between the 4th and ninth grade levels. The modules are on topics of interest to adults and provide information that adults want to know. The ten module topics cover consumer, health, and citizenship issues.

The learners can select a topic of interest from the list below and begin to improve their reading skills.

Tetanus
Heart Attacks
Eating Right To Avoid Health Problems
Buying A Used Car
Saving Money With Generic Drugs
Buying Good Nutrition With Fewer Dollars
Buy Now, Pay Later - Using Credit
Rights And Responsibilities Of Renters

Say What You Think By Voting
What You Should Do if You Are In A Car Accident

When the learners' use the READY Course, they choose one of the ten module topics and start the program. The program begins with the setting of goals for reading an article about the selected topic. The learners then read the article and explore the vocabulary in the text. When they know the vocabulary, they can begin to master and apply higher level reading comprehension skills to the article. As the learners work through additional READY COURSE modules their ability to use new skills becomes automatic and part of the normal reading process.

Below, is a paragraph from the module "Buy Now, Pay Later - Using Credit". The vocabulary words are highlighted in bold print. The learners can use the mouse to click on any bold word to see a pronunciation key and the definition of the word.

THE READY COURSE - Buy Now, Pay Later   Paragraph 1 of 4

The use of a credit can help you buy things you need when you don't have enough cash. You can use a *charge card* to take *advantage* of special sales. *Personal installment* loans can make it *possible* for you to *purchase* a car or *appliance*. But when using credit, you must be careful not to get in the trap of buying more than you can *afford* and *borrowing* more than you can pay back.

     charge card (charj kard) a <u>charge card</u> is a small
     plastic card that allows you to buy something on
     credit and pay for it at a later time.

                                            SOUND?
     Click on this line to see the next paragraph   ICON

On the sample screen above, the phrase *charge card* has been chosen. The pronunciation key and definition are in a box below the paragraph. In the lower right hand corner is an icon that the learner can click on to hear the pronunciation and definition of the vocabulary word. Learners can choose to hear the pronunciation and definition of at least 30 words for each module. Audio help is also available for the directions to each of the exercises.

Once the learners understand the content of the article, they use the content to learn more advanced skills such as scanning, organizing, summarizing, etc. Each READY Course module contains seven lessons that provide instruction and activities to help the learner increase the following reading skills:
1. Setting Reading Goals
2. Vocabulary Development
3. Scanning for Information
4. Making First Level Inferences
5.Organizing Information
6. Summarizing Information
7. Answering Comprehension Questions

The READY Course has a built-in course management and tracking system. Up to 105 learners can be enrolled in each module. The tracking system records all student scores by module and by activity as well as learner time on task. This tracking system also keeps progress records for the learners. It lets each learner know which modules and activities have been completed. The tracking system provides learner empowerment and control over the movement through the material.

One of the most important features of the READY Course is the use of digital audio. Digital audio can produce extremely high quality sound. A human voice is recorded and digitized so it can be stored as part of the computer program. The program is then stored on a CD-ROM disk. Students using the READY Course hear the audio through headphones plugged into the CD-ROM drive. Within the course, the audio is available when the four paragraphs of text are first introduced and again during the vocabulary development exercises. Audio is also available if the student chooses to hear the directions for each of the exercises spoken. It's important to note that the audio is an optional feature for the students and does not slow the progress of those students who choose not to use it.

2 READY Course Evaluation Study

2.1 Method

Participants in the General Education Development Diploma (GED) program reading courses were randomly placed into experimental and control groups at each of the seven field test sites. The subjects in the experimental group were given reading instruction using the READY Course delivered on CD-ROM over a 12 week quarter of the academic year. The control group was given traditional Adult Basic Education instruction using conventional instructional materials such as workbooks, low level high interest reading materials, and conventional classroom instruction. Both groups were testing using the TABE (Reading Vocabulary and Reading Comprehension = Total Reading G.E. Score) before instruction began and posttested using the TABE at the end of the 12 week period.

2.2 Instruments
Tests of Adult Basic Education (TABE)

The community colleges which participated in this field test all use the CTB McGraw-Hill Tests of Adult Basic Education (TABE) to diagnose and place their students for instructional purposes. They also used TABE to determine student reading improvement by GE score.

The TABE was revised by CTB/McGraw-Hill in 1976. The test contains a Reading, Mathematics and Language section. The reading section of the TABE generates a Vocabulary score, Comprehension score, and a Total Reading score. The Total Reading score of that instrument was used in this study. The Reading portion of the TABE was developed from items used in the California Achievement Test (CAT) 1970 version. There are four overlapping levels and two forms of the test. These levels include "E" that tests GE from 2.6 through 4.9, "M" that tests GE from 4.6 through 6.9, "D" that tests GE from 6.6 through 8.9 and "A" that tests GE from 8.9 through 12.9. A Locator Test which includes 25 reading vocabulary and 25 mathematics items is used by the instructor to determine the best level of the TABE to administer for detailed GE score information using LEVEL E, M, D, or A of the complete battery. All levels and forms of the TABE include 30 Reading Vocabulary and 40 Reading Comprehension test items.

To generate the Total Reading Score used for this study, raw scores for the Reading Vocabulary and Reading Comprehension are totaled. A conversion chart used to generate GE and Scale Scores for Total Reading is provided in the TABE Technical Report (1987) on pages 131 - 135 based on the norms for Adult Basic Education Students.

According to the TABE Technical Report (1987), the test was developed and both face and content validity were determined. Procedures were used by the test developers to control for bias, and standard item analysis was used to select items for the four levels and two forms of the test. Additional validity information was generated by an attempt to determine the relationship between the TABE and General Education Development tests (GED). To determine this relationship, 678 subjects were given the TABE and the GED and correlations were run using univariate regression.

Reliability studies were conducted using target population groups of Adult Basic Education, Adult Offender, Juvenile Offender, and Vocational/ Technical students as subjects. The 8125 subjects were from 35 states and 95 different education institutions. For the evaluation of the READY Course, all comparisons were made to the Adult Basic Education (ABE) norm group. The ABE norm group was made up of the following ethnic composition.

Ethnic Group	Percent of Sample	
Others	37%	(Non black, Hispanic, or Asian)
Blacks	45%	
Hispanics	12%	
Asians	5%	

TOTAL	100%	

Reliability for the TABE was reported by CTB/McGraw Hill as .68 for the complete battery. Reliability information was reported for the Adult Basic Education norm group using all forms and levels of the TABE. The following information provides an overview of the reliability data provided in the TABE Technical Report (1987, pps. 41-42).

THE (TABE)
LEVEL FORM   KR 20 RELIABILITY   KR 20 RELIABILITY
               Reading VOCABULARY   Reading
COMPREHENSION

		KR 20 RELIABILITY Reading VOCABULARY	KR 20 RELIABILITY Reading
A	5	.87	.87
	6	.86	.89
D	5	.90	.92
	6	.89	.91
M	5	.91	.93
	6	.90	.92
E	5	.91	.92
	6	.91	.91

2.3 Sample

Subjects for this study were 488 Adult Basic Literacy Education (ABLE) student volunteers at seven community colleges in North Carolina. While only three community college sites provided aggregate data on the experimental READY Course, all seven sites provided comparable test scores from participants in traditional reading instruction programs.

Once control group data was returned to the evaluators, 250 control group subjects were selected to match the percentage composition of the experimental group (n = 238) on the basis of ethnicity and sex. This technique produced the following Experimental (E) and Control group (C) distributions for ethnicity: E Caucasians (37%), C Caucasians (38%); E Blacks (53%), C Blacks (53%); E Hispanics (5%), C Hispanics (5%); and E Asians (3%), C Asians (3%). At the same time, males constituted 44% of the E group and 44% of the C group, while females constituted 56% of the E group and 56% of the C group.

2.4 Results

Due to different site procedures among the seven community colleges providing test data, the only comparable reading data among all subjects were Grade Equivalent scores on the Total Reading portion of the TABE. Consequently, all test analyses were conducted on this data. Both Davis (1964) and Sax (1980) have demonstrated that GE scores represent ordinal level measurement. Consequently, appropriate nonparametric inferential statistics were used to evaluate the differential performance of experimental and control groups.

To determine the equivalence of Experimental and Control groups prior to the onset of reading instruction, a Mann-Whitney U test was conducted on pretest TABE Total Reading scores. The result, a z score conversion of U, was 1.62 (two tailed p < .11, ns), indicating that the two groups began their respective reading programs with equivalent reading abilities.

Table 1 contains descriptive statistics of Total Reading GE gains for the sample, subdivided by experimental and control groups, ethnicity, and sex. No sex differences were found. However, further analyses showed that the experimental group gained a mean GE Total Reading score of 20 months with a standard error of measurement of slightly less than one month. In comparison, the control group gained a mean of only six and one-half months, with a standard error of measurement also slightly less than one month. This difference was statistically significant according to the Mann-Whitney U test (z = 9.685, 2-tailed p < .0001).

To test for differences between ethnic groups, a Kruskal-Wallis 1-Way Analysis of Variance (H, suitable for ordinal data, is approximately distributed as chi-square for large samples) for the experimental and control group Total Reading gains was conducted. For the experimental group, there were significant differences by race with Hispanics and Blacks showing greater gains than Caucasians and Asians (chi-square corrected for ties = 13.30, p< .005, cases = 235, df = 2). In contrast, control group Hispanics and Asians out-gained Caucasians and Blacks (chi-square corrected for ties = 14.64, p< .003, cases = 247, df = 2).

It could be argued that experimental group gains could have been due to a relatively small number of students who, due to special motivation, completed more READY Course reading units than less motivated students. To test this line of reasoning, experimental subjects were divided into three groups as follows: (1) subjects who completed less than one unit, (2) those who completed between one and up to four units, and (3) those who completed more than four units. Using the Kruskal-Wallis procedure to compare the three groups, the greatest gains took place for those subjects who had completed more than one READY Course unit. The gains for those completing 1 to 3 READY Course units was about the same as for those who had completed 4 to 6 READY Course units, but these were significantly higher than for students completing less than one READY Course unit (chi-square = 22.69, df = 2, p < .0001).

Table 1 Experimental and Control Group Mean Grade Equivalent (GE) Gains for TABE Total Reading Scores

Sample	Experimental Mean	SD	N	Control Mean	SD	N
	1.84	1.46	227	.59	1.10	239
By race						
Caucasian	1.50	1.46	87	.70	1.19	95
Blacks	1.99	1.43	128	.48	1.05	132
Hispanic	2.65	1.36	12	1.03	.69	12
By sex						
Male	1.61	1.36	104	.63	1.28	109
Female	1.96	1.50	134	.66	1.13	140

2.5 Conclusions

1. There were significant differences in pre- to posttest G.E. score gains by race.
   - For the experimental group, Hispanics and Blacks showed the greatest gains.
   - For the control group, Blacks showed the lowest gain, while Hispanics and Asians as well as Caucasians showed the greatest gains.

Both methods of instruction produced significant gains overall. The READY Reading course seemed to influence the gains in the Black population more than

ABE reading instruction. Based on earlier findings (LSS STUDY) suggesting that blacks see themselves as auditory learners this is a logical conclusion.

2. There were significant differences in pre - posttest G.E. score gains by number of READY Reading units completed.

    - The greatest gains took place once the subjects in the experimental group had completed one to three READY units and continued as they completed four to six units. This would suggest that students will need to complete one or more READY unit before significant gains in G.E. scores can be expected.

3. There were significant differences in pre- posttest G.E. score gains by group.

    - The average G.E. score gain for the experimental group was 1.8 school years over the 12 week instructional period. The average G.E. score gain for the control group was .59 school years over the 12 week instructional period.

    Based on these findings, the READY Course seems to produce much greater G.E. reading score gains than traditional instruction.

4. There were significant differences in pre- to posttest G.E. score gains by both the <6th grade and >6th grade experimental group subjects based on their pretest TABE scores.

    - Of the 119 subjects in the experimental group who scored below the 6th grade reading level on the pretest, 112 showed gains from pre to posttesting.

    - Of the 119 subjects in the experimental group who scored above the 6th grade reading level on the pretest, 108 showed gains from pre to posttesting.

    Based on these findings, significant gains were made by both those who were below the 6th grade reading level on the pretest and also for those who were above the 6th grade reading level on the pretest. Pretest G.E. reading scores did not seem to be a factor in the pre- to posttest gains for READY Course students.

5. There were NO significant differences in G.E. reading score gains based on *age, sex,* or *previous reading instruction.*

3 The Future

In the overall comparison, interactive reading courseware in the form of the READY Course far surpassed traditional reading instruction when it came to gains of the low-literate adults in this study. Future research should focus on the use of such interactive technology to train or instruct other groups from preschool through graduate school.

    What might happen if the options opened to learners using the courseware were expanded to match their learning styles? What might happen if the learner control over the courseware were increased providing each learner with more instructional options? New training and instructional courseware should include self-assessment (Leclercq & Brogniez, 1990). It should be tied to individual learning style through flexible access based on style. New courseware for training and instruction might even include learning style diagnostic instruments. The answer to one over riding question should help us answer these other questions. That is "Do learners operate consistently in the way that they predict they will operate on learning style instruments?".

References

CTB/McGraw-Hill (1987) TABE Norms Book. Monterey, CA: CTB/McGraw-Hill

CTB/McGraw-Hill (1987) TABE Survey Form Examiner's Manual. Monterey, CA: CTB/ McGraw-Hill

CTB/McGraw-Hill (1987) TABE Technical Report. Monterey, CA: CTB/McGraw-Hill

Davis, Frederick B. (1964) Educational Measurements and Their Interpretation. Belmont, CA: Wadsworth

Gretes, J.A., Green, M., & Songer, T. (1992) Field Testing the READY Course: Interactive Instruction with Digitized Audio is Making a Difference. Proceedings of the Ninth International Conference on Technology and Education, Paris

Gretes, J.A., & Songer, T. (1990) The learning style survey (LSS), An interactive videodisc instrument: An instrument validation study. In Proceedings of the Seventh International Conference on Technology and Education, Brussels, vol. 2, 435-437

Gretes, J.A., & Songer, T. (1989) Validation of the learning style survey: An interactive videodisc instrument. Educational and Psychological Measurement 49, 235-241

Gretes, J.A., & Songer, T. (1988) Central Piedmont community college learning style survey (LSS) interactive videodisc project instrument validation report. Charlotte, NC: The READY Project

Hill, J.E. (1970). Cognitive style as an education science. Bloomfield Hills, MI: Oakland Community College Press

Hill, J.E. (1980) The educational sciences: A conceptual framework. West Bloomfield, MI: Hill Educational Sciences Research Foundation

Hill, J.E. & Nunney, D.N. (1974) Personalizing educational programs utilizing cognitive style mapping. Bloomfield Hills, MI: Oakland Community College Press

Leclercq, D. & Brogniez, P. (1990) A fresh look on confidence making. In Proceedings of the Seventh International Conference on Technology and Education, Brussels, vol. 1, 646 -649

Leclercq, D. & Boskin, A. (1990) Note taking behaviors studied with the help of hypermedia. In Proceedings of the Seventh International Conference on Technology and Education, Brussels, vol. 2, 16-19

Leclercq, D. & Pierret, B. (1989) A computerized open learning environment to study interpersonal variations in learning styles: DELIN. In Proceedings of the Sixth International Conference on Technology and Education, Orlando, FL, vol. 2, 268-272

Sax, Gilbert. (1980) Principles of Educational and Psychological Measurement and Evaluation. Belmont, CA: Wadsworth

Siegel, Sidney. (1956) Non-Parametric Statistics for the Behavioral Sciences. New York: McGraw-Hill

Participants and Contributors

Prof. Antonio Bartolome
University of Barcelona
Amadeu Vives, 8
08320 El Masnou, Barcelona, Spain

Prof. James Bruno
University of California
Graduate School of Education
405 Hilgard Avenue
Los Angeles, CA 90024-1321, USA

Mr. Philippe De Brogniez
University of Liège
Boulevard du Rectorat, 5 Bât. B32
Sart Tilman, 4000 Liège 1, Belgium

Mrs. Alberta Della Piane
CNITE Roma, Via Taro, 35
00185 Roma, Italy

Prof. Jean-François Denef
Catholic University of Louvain
Faculté de Médecine
Laboratoire d'Histologie
UCL 5229, 1120 Brussels, Belgium

Mrs. Brigitte Denis
University of Liège
Boulevard du Rectorat, 5 Bât. B32
Sart Tilman, 4000 Liège 1, Belgium

Prof. Arie Dirkzwager
Bussum, Huizerweg, 62
1402 AE Bussum, Netherlands

Prof. Bernard Dumont
University of Paris VII
Place Jussieu, 2
U.F. de Didactique des Disciplines
75251 Paris Cedex 05, France

Prof. Georges Eisendrath
Free University of Brussels
Educo, Pleinlaan 2
1050 Brussels, Belgium

Mr. Jean-Marc Fabre
University of Provence
Avenue Robert Schuman, 29
13621 Aix-en-Provence,
Cedex 1, France

Mrs. Anna Gammaldi
FORMEZ - Roma
Via Rubicone, 11
Roma, Italy

Mr. Jean-Luc Gilles
University of Liège
Boulevard du Rectorat, 5 Bât. B32
Sart Tilman, 4000 Liège 1, Belgium

Prof. John Alexander Gretes
University of North Carolina
College of Education
Charlotte, NC 28223, USA

Prof. Darwin P. Hunt
Human Performance
Enhancement, Inc.
Executive Center II
345 North Water Street
Las Cruces, NM 88001, USA

Mr. Gideon Keren
Free University of Amsterdam
Department of Psychology
De Boelelaan, 1111
1081 HV Amsterdam
The Netherlands

Prof. Gilbert De Landsheere
University of Liège
Boulevard du Rectorat, 5 Bât. B32
Sart Tilman, 4000 Liège 1, Belgium

Prof. Dieudonné Leclercq
University of Liège
Boulevard du Rectorat, 5 Bât. B32
Sart Tilman, 4000 Liège 1, Belgium

Mrs. Renata Picco
Ministero della Publica Istruzione
Centro Europeo Dell Educazione
Villa Falconieri
00044 Frascati (Roma), Italy

Mr. Alexander Renkl
Universität München
Institut für Empirische Pädagogik
und Pädagogische Psychologie
Leopoldstraße 13
80802 München, Germany

Mrs. Cristina Salgado
LNETI Lisbonne
Azinhaga Dos Lameiros A Estrada
Do Paco Do Lumiar
1699 Lisboa, Portugal

Mr. Charles Schmit
University of Liège
STE-Formations Bât. C1
Rue A. Stévart n° 2
4000 Liège, Belgium

Emir Hamvasy Shuford
The Knowledge Group
P.O. Box 25102
Dallas, TX 75225, USA

Cmdt J.-P. Straetmans
Technical School of the Air-Force
Caserne Colonel Aviateur Renson
3800 Sint-Truiden, Belgium

Prof. Ata Tezbasaran
Student Selection and Placement
Center of Ankara OSYM
06538 Ankara, Turkey

Dr. Jelle Van Lenthe
University of Groningen
Dept. of Statistics and Measurement
Theory
Grote Kruisstraat 2/1
9712 TS Groningen
The Netherlands

Prof. Benedetto Vertecchi
University of Roma "La Sapienza"
Via del Castro Pretorio, 20
00185 Roma, Italy

Mrs. Elleke Verwaijen
CIBB
Pettelaarpark 1 's-Hertogenbosch
Postbus 1585
5200 BP's-Hertogenbosch
The Netherlands

Mr. Romain Zeiliger
FNRS - IRPEACS Lyon
Chemin des Mouilles, BP 167
69131 Ecully, France

NATO ASI Series F

Including Special Programmes on Sensory Systems for Robotic Control (ROB) and on Advanced Educational Technology (AET)

Vol. 1: Issues in Acoustic Signal - Image Processing and Recognition. Edited by C. H. Chen. VIII, 333 pages. 1983.

Vol. 2: Image Sequence Processing and Dynamic Scene Analysis. Edited by T. S. Huang. IX, 749 pages. 1983.

Vol. 3: Electronic Systems Effectiveness and Life Cycle Costing. Edited by J. K. Skwirzynski. XVII, 732 pages. 1983.

Vol. 4: Pictorial Data Analysis. Edited by R. M. Haralick. VIII, 468 pages. 1983.

Vol. 5: International Calibration Study of Traffic Conflict Techniques. Edited by E. Asmussen. VII, 229 pages. 1984.

Vol. 6: Information Technology and the Computer Network. Edited by K. G. Beauchamp. VIII, 271 pages. 1984.

Vol. 7: High-Speed Computation. Edited by J. S. Kowalik. IX, 441 pages. 1984.

Vol. 8: Program Transformation and Programming Environments. Report on a Workshop directed by F. L. Bauer and H. Remus. Edited by P. Pepper. XIV, 378 pages. 1984.

Vol. 9: Computer Aided Analysis and Optimization of Mechanical System Dynamics. Edited by E. J. Haug. XXII, 700 pages. 1984.

Vol. 10: Simulation and Model-Based Methodologies: An Integrative View. Edited by T. I. Ören, B. P. Zeigler, M. S. Elzas. XIII, 651 pages. 1984.

Vol. 11: Robotics and Artificial Intelligence. Edited by M. Brady, L. A. Gerhardt, H. F. Davidson. XVII, 693 pages. 1984.

Vol. 12: Combinatorial Algorithms on Words. Edited by A. Apostolico, Z. Galil. VIII, 361 pages. 1985.

Vol. 13: Logics and Models of Concurrent Systems. Edited by K. R. Apt. VIII, 498 pages. 1985.

Vol. 14: Control Flow and Data Flow: Concepts of Distributed Programming. Edited by M. Broy. VIII, 525 pages. 1985.

Vol. 15: Computational Mathematical Programming. Edited by K. Schittkowski. VIII, 451 pages. 1985.

Vol. 16: New Systems and Architectures for Automatic Speech Recognition and Synthesis. Edited by R. De Mori, C.Y. Suen. XIII, 630 pages. 1985.

Vol. 17: Fundamental Algorithms for Computer Graphics. Edited by R. A. Earnshaw. XVI, 1042 pages. 1985.

Vol. 18: Computer Architectures for Spatially Distributed Data. Edited by H. Freeman and G. G. Pieroni. VIII, 391 pages. 1985.

Vol. 19: Pictorial Information Systems in Medicine. Edited by K. H. Höhne. XII, 525 pages. 1986.

Vol. 20: Disordered Systems and Biological Organization. Edited by E. Bienenstock, F. Fogelman Soulié, G. Weisbuch. XXI, 405 pages.1986

Vol. 21: Intelligent Decision Support in Process Environments. Edited by E. Hollnagel, G. Mancini, D. D. Woods. XV, 524 pages. 1986.

NATO ASI Series F

NATO ASI Series F

NATO ASI Series F

Including Special Programmes on Sensory Systems for Robotic Control (ROB) and on Advanced Educational Technology (AET)

NATO ASI Series F

Including Special Programmes on Sensory Systems for Robotic Control (ROB) and on Advanced Educational Technology (AET)

Vol. 83: Active Perception and Robot Vision. Edited by A. K. Sood and H. Wechsler. IX, 756 pages. 1992.

Vol. 84: Computer-Based Learning Environments and Problem Solving. Edited by E. De Corte, M. C. Linn, H. Mandl, and L. Verschaffel. XVI, 488 pages. 1992. *(AET)*

Vol. 85: Adaptive Learning Environments. Foundations and Frontiers. Edited by M. Jones and P. H. Winne. VIII, 408 pages. 1992. *(AET)*

Vol. 86: Intelligent Learning Environments and Knowledge Acquisition in Physics. Edited by A. Tiberghien and H. Mandl. VIII, 285 pages. 1992. *(AET)*

Vol. 87: Cognitive Modelling and Interactive Environments. With demo diskettes (Apple and IBM compatible). Edited by F. L. Engel, D. G. Bouwhuis, T. Bösser, and G. d'Ydewalle. IX, 311 pages. 1992. *(AET)*

Vol. 88: Programming and Mathematical Method. Edited by M. Broy. VIII, 428 pages. 1992.

Vol. 89: Mathematical Problem Solving and New Information Technologies. Edited by J. P. Ponte, J. F. Matos, J. M. Matos, and D. Fernandes. XV, 346 pages. 1992. *(AET)*

Vol. 90: Collaborative Learning Through Computer Conferencing. Edited by A. R. Kaye. X, 260 pages. 1992. *(AET)*

Vol. 91: New Directions for Intelligent Tutoring Systems. Edited by E. Costa. X, 296 pages. 1992. *(AET)*

Vol. 92: Hypermedia Courseware: Structures of Communication and Intelligent Help. Edited by A. Oliveira. X, 241 pages. 1992. *(AET)*

Vol. 93: Interactive Multimedia Learning Environments. Human Factors and Technical Considerations on Design Issues. Edited by M. Giardina. VIII, 254 pages. 1992. *(AET)*

Vol. 94: Logic and Algebra of Specification. Edited by F. L. Bauer, W. Brauer, and H. Schwichtenberg. VII, 442 pages. 1993.

Vol. 95: Comprehensive Systems Design: A New Educational Technology. Edited by C. M. Reigeluth, B. H. Banathy, and J. R. Olson. IX, 437 pages. 1993. *(AET)*

Vol. 96: New Directions in Educational Technology. Edited by E. Scanlon and T. O'Shea. VIII, 251 pages. 1992. *(AET)*

Vol. 97: Advanced Models of Cognition for Medical Training and Practice. Edited by D. A. Evans and V. L. Patel. XI, 372 pages. 1992. *(AET)*

Vol. 98: Medical Images: Formation, Handling and Evaluation. Edited by A. E. Todd-Pokropek and M. A. Viergever. IX, 700 pages. 1992.

Vol. 99: Multisensor Fusion for Computer Vision. Edited by J. K. Aggarwal. XI, 456 pages. 1993. *(ROB)*

Vol. 100: Communication from an Artificial Intelligence Perspective. Theoretical and Applied Issues. Edited by A. Ortony, J. Slack and O. Stock. XII, 260 pages. 1992.

Vol. 101: Recent Developments in Decision Support Systems. Edited by C. W. Holsapple and A. B. Whinston. XI, 618 pages. 1993.

NATO ASI Series F

Including Special Programmes on Sensory Systems for Robotic Control (ROB) and on Advanced Educational Technology (AET)

The manufacturer's authorised representative in the EU is Springer
Nature Customer Service Centre GmbH, Europaplatz 3, 69115 Heidelberg,
Germany. If you have any concerns regarding our products, please
contact ProductSafety@springernature.com

Printed and bound by CPI Group (UK) Ltd, Croydon, CR0 4YY

29/04/2026

02099460-0017